J.D. Whitney

Earthquakes, Volcanoes and Mountain-Building

Three Articles Published in the Northern American Review 1869-1871

J.D. Whitney

Earthquakes, Volcanoes and Mountain-Building
Three Articles Published in the Northern American Review 1869-1871

ISBN/EAN: 9783743406797

Manufactured in Europe, USA, Canada, Australia, Japa

Cover: Foto ©berggeist007 / pixelio.de

Manufactured and distributed by brebook publishing software
(www.brebook.com)

J.D. Whitney

Earthquakes, Volcanoes and Mountain-Building

EARTHQUAKES, VOLCANOES,

AND

MOUNTAIN-BUILDING.

THREE ARTICLES PUBLISHED IN THE NORTH AMERICAN REVIEW,
1869–1871.

By J. D. WHITNEY.

―――――――

UNIVERSITY PRESS, CAMBRIDGE.
1871.

CONTENTS.

NOTE. — The first of these articles appeared in the North American Review for April, 1869, Vol. CVIII. page 578 ; the second, July, 1869, Vol. CIX. p. 231 ; the third, October, 1871, Vol. CXIII. page 235. A few extra copies of each were struck off and are here placed together for convenient distribution to friends and those who may be specially interested in the subject of volcanism.

J. D. W.

CAMBRIDGE, October, 1871.

EARTHQUAKES.

1. *The First Principles of Observational Seismology.* By R. MALLET. 2 vols. Royal 8vo. London. 1862.
2. *Untersuchungen über das Phänomen der Erdbeben in der Schweiz.* Von G. H. OTTO VOLGER. 3 Theile. Gotha. 1857.
3. *Volcanoes and Earthquakes.* By MM. ZURCHER and MARGOLLÉ. 8vo. Philadelphia and London. 1869.

THE titles placed at the head of this article indicate three as characteristic books as could be selected from among the mass of publications devoted either to earthquakes alone or to earthquakes and volcanoes conjointly. In the last one on the list we have a fair specimen of a class of books which are becoming quite common, which mostly originate in France, are translated in England, and are reprinted here, and which, while pretending to be scientific, are, in reality, as far from having any claim to that character as possible. The principle on which these books are got up seems to be this: A number of showily designed and elegantly engraved wood-cuts are manufactured, and then some scientific penny-a-liner is hired to put together a text to match the pictures, no time being allowed for doing the work properly, even if the person selected were competent, — which is rarely the case, — the dominating idea being, evidently, to produce something which a not very critical public shall be tempted into buying, on account of the beauty of its mechanical execution, and with the incidental advantage of getting something scientific into the bargain.

The materials for the illustrations and text of such books are taken right and left without acknowledgment, the one caricatured and the other "popularized," — that is to say, enormously exaggerated or misrepresented, partly through ignorance, but chiefly through a desire to produce a sensational

1

effect. The result is even worse than that produced by the
modern sensational novel ; for the latter is read, thrown away
and forgotten, while the pseudo-scientific and elegantly illustrated volume is carefully laid away in the book-case, and
referred to as a standard authority, and most certainly added
to the dead weight of every public library, crowding out that
which is really valuable in the same department, and which is
overlooked, perhaps because it is a little old, because its exterior is not attractive, or because its appearance has not been
heralded by a publisher's fanfare.

There could not be a better instance selected, as a text on
which to preach a sermon, *à propos* of this style of illustrated
works, than that furnished us by this book of MM. Zurcher and
Margollé, whoever they may be. The illustrations are showy,
and, as far as the engraving is concerned, well executed,
though badly printed in the English edition from the pur-
chased electrotypes, while, in the original, that branch of the
mechanical execution was undoubtedly carefully attended to.
But let any one conversant with the subject of which the volume treats examine the illustrations, and he will see at once
that the drawings were made by persons entirely ignorant of
what they were attempting to represent. Thus, in the views
opposite pages 10 and 34, an attempt is made to show the
phenomena of violent eruptions of Etna and Vesuvius. Now,
if there is any characteristic feature of these eruptions, it is the
straightness of the column in which the projected material as-
cends until it reaches its highest point. In these drawings, on
the contrary, apparently in order to add the curve of beauty
and grace to the picture, the column, in both cases, is made to
ascend in an elegantly waving line, as untrue to Nature as possible. Again, every one is familiar with the sketch of Cotopaxi,
given by Humboldt, as an illustration of a beautifully regular
volcanic cone, and which has figured in hundreds of books
for the last fifty years, — notably in our school geographies.
Humboldt, in his sketch, misled by the invariable tendency of
the eye to exaggerate the slope of mountains, represented the
inclination of the sides of the cone as 48°, while the photo-
graphs show that in reality this angle is no more than 28° or
29°, the greatest inclination, just at the summit, being only 32°.

Now, on the wood-cut in MM. Zurcher and Margollé's book, the slope of all the snow-covered part of the volcano is given as 55°, while the effect of the whole is very much like what one might imagine would be produced by a stove-funnel perched on the summit of a big boulder. This is the character of the illustrations throughout; there is one, however, which surpasses all the others in its ludicrous absurdity,—representing a great number of Calabrian peasants in the act of being swallowed by earthquake chasms, the whole style of the thing being well suited to the pages of a comic almanac, perhaps, but certainly not to those of a scientific work. Of the text of this book it may be said, without hesitation, that it is fit to go with the illustrations. Let a single suggestion be quoted from it, to show how that which is unsound in theory, but at the same time brilliant and peculiarly French, is mixed with something supposed to be popularly and economically interesting, — the idea being to convey the impression that science has its practical as well as its abstruse side. M. Élie de Beaumont, a distinguished French geologist, has devoted much time to tracing out on the earth a regular geometric arrangement, with which he thinks the lines of upheaval of mountain chains may be found to coincide, and which he calls " a pentagonal network." The idea is ingenious, and has been elaborately wrought out by its author, but accepted by few of the leading geologists at the present day. Our authors, however, make both themselves and the pentagonal network ridiculous, by advising that it should be used as a guide in boring for springs of petroleum. They even trace an imaginary line from Iceland straight to Oil Creek, " places remarkable for their bituminous emanations," as a guide to " oil prospectors." What a pity this brilliant idea had not been suggested before the collapse of the great bubble! One might then have had the " Great Consolidated Pentagonal Network Petroleum Company " to add to the list of other remarkable things in that line. All that is said of the volcanoes and volcanic rocks of our own country in MM. Zurcher and Margollé's book is equally curious, as an exhibition of entire ignorance of our geography and geology. There are just about as many misstatements as there are lines in the two pages devoted to North American volcanoes.

Herr Volger's book is as thoroughly German as that just noticed is French. This author, living at a distance from any region of volcanoes and great earthquakes, but in one where moderate shocks are frequent, and having a strong propensity to look at natural phenomena in what may be called " the small way," has evolved a theory of earthquakes from the depths of his moral consciousness, and then endeavored to bolster it up by collecting great numbers of facts, also of the small kind, entirely ignoring the greater facts, to which his smaller ones are as the ripple on the surface of the ocean in a gentle breeze to the great tidal wave which encircles the whole globe in its motion, and stirs the waters to the very profoundest depths. The book, however, quite different from that of MM. Zurcher and Margollé, is valuable as presenting the extreme views of the school to which the author belongs, and as extremely ingenious in its defence of them, although thoroughly wrong in its fundamental ideas, — as much so as one would be who should endeavor to work out the comparative anatomy of the elephant by a microscopic examination of the pimples on his hide.

Mr. Mallet's book is as different from either of the others as possible. In order to make its character intelligible, it will be necessary to give some idea of his previous publications, and of those of other really scientific investigators in the same line of research, and to show how and when this branch of geological science acquired a right to the special name it now bears, that of SEISMOLOGY, a term derived directly from the Greek, and signifying the Science of Earthquakes.

The phenomena of volcanic and earthquake action, inseparably connected in the popular mind, and not easily disentangled from each other by the scientific, must necessarily engross a large share of thought in regions where they are frequently manifested, and especially at the time when such manifestations are peculiarly violent or destructive. As no exhibition of the forces of Nature is so sublime as that of the volcano, or so fearful in its consequences as that of the earthquake, it was natural not only that allusions to these phenomena should be found in the oldest writings of all nations inhabiting regions liable to such visitations, but that their very cosmogonies should be profoundly affected by these workings of unseen forces. Hence, in all the sacred writings of the nations inhabiting the vicinity of that cradle of civilization, the Mediterranean, a region liable to earthquakes, and well provided with volcanoes, we find a substratum of belief in occasional conflagrations and deluges by which the gods were wont to arrest the career of human wickedness, and to sweep off from the face of the earth its inhabitants, in order to make way for a new and improved creation. Such ideas pervaded the Egyptian, the Hindoo, the Hebrew, the Arabian, and the Greek mythology ; and vestiges of the same are found in the earthquake-shaken regions of South America ; while we are not aware that any trace of them can be discovered in the cosmogony of the North American Indians, dwelling in a region but little liable even to slight earthquake shocks, and entirely free from volcanoes.

In consequence of the effect which violent earthquake shocks produce on the material progress of the countries subject to them, and of their direct relations to the welfare of the human race, it is evident that these and kindred events stand in a different relation to history from the ordinary phenomena of geology. Those operations of Nature which proceed slowly and quietly, without destruction of life and property, are not so calculated to excite immediate and universal attention as those which are accompanied by devastating effects over vast regions. Yet the former class may in reality be of as much importance in modifying the surface of the globe, and may finally bring about results as momentous, as any immediately following great earthquake shocks or volcanic eruptions. Thus, slow upheavals or depressions of large areas of land do really produce changes profoundly affecting the welfare of great numbers of people ; but these changes take place so slowly that they are prepared for beforehand, and their effect is spread over very long periods, and is not strikingly perceptible at any one moment. That such changes were occurring in past times is clearly demonstrated, and there is every reason to suppose that they are going on now. Indeed, it is not difficult for the geologist to point to the very regions where slowly, but surely, the ground is sinking over large areas, and where alterations in the distribution of sea and land will in time have accumulated sufficiently to produce materially important results in relation to vegetation, animal life, and the development of the human race.

In view of the above, it may well be supposed that the number of volumes devoted to descriptions of the phenomena of earthquakes and volcanoes, in all languages, is very great. M. Alexis Perrey has given a list of eighteen hundred and thirty-seven works on earthquakes published up to 1856 ; but he admits that it is not complete. Those relating to volcanoes and volcanic phenomena are undoubtedly more numerous, but no one appears to have set about making a full catalogue of them. On the subject of Vesuvius alone, the list of volumes and scientific papers — many of the latter much more important than some of the separate works — would embrace many hundred titles, perhaps thousands. Most of these works are purely descriptive of the incidents observed, as the number of persons killed, or of buildings thrown down ; they do not attempt to philosophize or generalize, or, if they do, it is in a moral or religious strain rather than a scientific one. The descriptions of volcanic phenomena are, on the whole, much more satisfactory than those of earthquakes, as will be easily understood. The former are more at the command of the observer, are longer in duration, and less appalling in their results. Earthquakes, on the other hand, at least those of magnitude, come unexpectedly, give no indication of their probable violence or duration, and so unfit the mind for calm observation that it is only by previously arranged automatic machinery that we can expect to get accurate information as to those points in regard to which the palpable mementos of their occurrence are not left after the shock has subsided. Hence it is only of late years that earthquake phenomena have begun to be thoroughly investigated, and that the science of Seismology has taken a position among those branches of research in which accuracy of statement is expected.

Since history has preserved more or less complete records of earthquake shocks, in various parts of the world, from very early times, it would naturally be inferred that light might be thrown on some of the obscure points in seismological science by carefully collecting and tabulating all that is scattered through innumerable published volumes, in regard to the time, place, and extent of earthquake phenomena, and that this would be the necessary preparation for any thorough working-up of the subject. That such a working-up was desirable was evident; for, although most geologists, following Humboldt and Buch, were pretty clearly in accord with regard to the general cause of volcanic eruptions and earthquakes, yet much remained to be done to make the *modus operandi* of the internal forces more clear, and especially to furnish material for combating the views of a small class of geologists who persistently refuse to adopt the views formulated by Humboldt, and desire, on the other hand, to refer all these phenomena, as far as possible, to local causes, as will be explained farther on.

The paper read by Humboldt, in 1823, before the Berlin Academy, " On the Structure and Mode of Action of Volcanoes in Different Parts of the Globe," contained the first truly philosophical discussion of this subject ; and the ideas with regard to the origin of volcanoes and earthquakes, then first rather vaguely announced, and afterwards more clearly formulated in the " Cosmos " (1845), have been the guiding thread by which most investigators in this branch of science have endeavored to work their way through the labyrinth of difficulties in which they have found themselves involved. The grand generalization of Humboldt, by which the whole subject of the theory of these phenomena was summed up in a few words in the " Cosmos," is as follows : " In a comprehensive view of Nature's operations, all these [namely, the phenomena of volcanism and earthquake action] may be fused into the one simple idea of *the reaction of the interior of the earth upon its exterior*." Here was simply and concisely enunciated the guiding principle of modern structural geology, and one which, by its subsequent connection with the nebular hypothesis, has become more and more generally recognized as the cornerstone on which the science rests. It is certainly true, that, if this theory be not adopted, there is no central idea in the science, nothing about which it can crystallize, and that the whole assemblage of facts so laboriously collected in physical geology is without anything to compact it into one harmonious whole.

Among those writers who have devoted considerable time to the investigation of earthquake phenomena, besides Humboldt, are: Von Hoff, in whose History of the Changes which have taken Place in the Condition of the Earth's Surface (Gotha, 1822–41) much information is to be found; Friedrich Hoffmann, in a variety of elaborate papers, and especially in his posthumous works, chiefly published between 1831 and 1839; F. C. Kries, whose work on the Causes of Earthquakes was crowned and published by the Dutch Academy in 1820. Peter Merian also made an elaborate investigation of the earthquakes occurring at Basle; Arago published several valuable papers on volcanoes and earthquakes, from 1820 to 1824; and Gay-Lussac contributed an important paper, in 1823, on the theory of volcanoes, in which was the first definite recognition of the vibratory character of earthquake motions. In 1846, Mr. Robert Mallet, of Dublin, published his first paper on the Dynamics of Earthquakes, in the Transactions of the Royal Irish Academy. In 1847, Mr. W. Hopkins furnished his Report on the Theories of Elevation and Earthquakes to the British Association, a paper which has been much quoted and used by various writers on geological subjects. M. Alexis Perrey is, however, undoubtedly the most voluminous writer on earthquakes; his papers and publications are scattered through a great number of the Journals and Transactions of learned societies, — chiefly those of the French and Belgian Academies, — and bear date from 1841 to 1868; the latest of which we have learned being a statistical account of the earthquakes of Alaska. In the British Association Report for 1858 will be found a list of M. Perrey's publications, from the earliest down to 1858, including fifty-nine titles. His library was recently offered for sale, and was shown by the catalogue to contain four thousand and fifteen works on the two subjects of earthquakes and volcanoes.

Elaborate and valuable as M. Perrey's papers are, especially to those working in this department of science, they are chiefly statistical in their nature, and cannot be compared for scope and general ability with those of Mr. Mallet, the labors of the last-named seismologist being not only those of a compiler, but also of an original experimenter and observer in this field.

Mr. Mallet's results have been laid before the public chiefly in the form of Reports to the British Association, appearing in the volumes for 1850, 1851, 1852, 1853, 1854, and 1858. His principal separate publication is the one cited at the head of this article.

In the Reports of the British Association Mr. Mallet gives a catalogue of all recorded earthquakes from 1606 B. C. to 1842; but for the discussion of the subject in his last Report (1858) he uses the tabular statements of M. Perrey, thus supplementing his own work by that of M. Perrey, for the years 1843–50. All the catalogues up to that time gave, as a basis for induction, more or less precise information in regard to between six and seven thousand earthquakes. It is easy to see, that, the farther we go back in time, the more imperfect the records of earthquakes, as well as of all other physical events, will be found to be. As Mr. Mallet remarks, in speaking of the curves he has drawn, illustrating the frequency of recorded earthquakes during the different centuries: · " Our chrono-seismic curve is, in fact, not only a record of earthquakes, but a record of the advance of human enterprise, travel, and observation." Thus, for the years 1700 to 1400 B. C. there are a few scattered facts; then, from 1400 to 900, nearly five hundred years of perfect blank; followed again, with a few exceptions, by another blank from 800 to 600 B. C. Indeed, the only record of any value for scientific analysis commences about 500 B. C. Since that time, the epochs of the invention of printing and of the Reformation are clearly marked in the expansions of the curves; while the discovery of America, the voyage to India around the Cape of Good Hope, and the vast increase in the commercial intercourse of the world consequent thereon, are also perfectly recognizable in the rapid accumulation of data, and the sudden swelling of the curve of frequency. While only six or seven thousand earthquakes have been tabulated for all time down to 1850, a German author, Dr. K. E. Kluge, was able to obtain records of four thousand six hundred and twenty as occurring between the years 1850 and 1857, inclusive.

We will now endeavor to present, with a few additions of our own, the most important results obtained by the various authors specified, in working over the great mass of statistical information which has been accumulated.

Earthquakes may be considered, *first*, with reference to their geographical distribution, or the position which seismic areas occupy on the earth's surface with reference to each other, to the great features of the earth's surface, and to the position of areas where kindred phenomena — as, for instance, volcanic eruptions — are manifested ; *second*, in their relations of time, or with reference to their occurrence, as connected, synchronously or otherwise, with changes of the seasons, or as recurring in cycles, or as influenced by the position of the heavenly bodies, especially the moon or the sun ; and, *lastly*, as connected with movements or conditions of the atmosphere, or with electrical and magnetic disturbances.

Let us examine, first, what deductions can be drawn from the geographical position of earthquake areas.

There are several " seismographic maps," showing the geographical distribution of earthquake regions ; among these, the best and latest is that accompanying Mallet's Fourth Report to the British Association, and from which the others do not differ much in the general character of the results shown. The first impression produced by looking at any of these should be rather one of alarm ; for nearly the whole of the inhabited and habitable earth appears to be shaded with the various tints implying a greater or less liability to earthquake shocks. It would, indeed, seem, at first sight, as if only those regions were left uncolored in regard to which no information could be obtained. Thus, all of Europe is more or less deeply colored, except a part of Central and Northeastern Russia,— and nearly all of Asia, except the extreme northern portion of Siberia, and the country drained by the Amoor River. On the other hand, almost the whole of Africa and Australia is left blank, as well as the extreme northern portion of North America, and all of South America east of the eastern base of the Andes and south of a narrow belt extending around from New Granada to French Guiana. It is evident that portions of the areas thus omitted in the distribution of the earthquake tint must have been left blank on account of the absence of information in regard to their seismic character. Indeed, it might be asked whether there is any part of the world where earth-quakes do not occur. To this question it would probably be safe to reply, that there is no region thickly inhabited by a civilized people, and where consequently there is a pretty complete record of what has happened for a considerable period back, in which there are not occasional slight manifestations of seismic energy. But it is pretty certain, on the other hand, that regions so well known as Brazil, or as some portions of the coast of Africa, could not be much troubled by earthquakes without some information having been gathered in regard to them by the many travellers who have visited those countries. The fact that any area is left uncolored in Mr. Mallet's map is a strong reason for believing that it is probably not liable to severe shocks. Leaving Africa and Australia out of the question, as too little known to allow of positive statements being made concerning them, it may be said that there is hardly any portion of the habitable globe which is not occasionally shaken, but that Eastern South America comes nearest to a desirable state of security in this respect. Most of British North America is also very firm in its position as an integral part of the crust, but is not likely ever to be very thickly inhabited.

On close examination of Mr. Mallet's map, we see upon it three tints of color, intended to distinguish, as we learn from the accompanying report, the relative intensity of the shocks occurring in the regions designated by them. The deepest tint indicates great earthquakes; the middle tint, those of mean intensity; the lightest color, minor shocks. By "great earthquakes" are meant those in which, over large areas, numerous cities or towns are overthrown, persons killed, rocky masses dislocated, etc.; under the head of earthquakes of mean intensity are included such as were felt over a wide area, but which were not severe enough to have a very destructive effect, and were not attended with much loss of life. The third class embraces those slight tremors of the surface which do not produce any serious destruction or commotion, and which leave but few, if any, traces of their occurrence.

It will of course be impossible, in these pages, to enter into any minute discussion of the distribution of the different bands of seismic energy; but some general idea of their geographical range can be given. Let us examine, first, the position of the patches of deepest tint, — those indicating the occurrence of " great earthquakes." We see at once that the area thus colored is, to a very large extent, coincident with that of the greatest displays of active volcanic forces. As the whole Pacific coast of America and the islands of the coast of Asia are the scene of volcanic phenomena on the grandest scale, so, too, the darkest tint of color follows the coast of the Pacific Ocean around, indicating great earthquakes along nearly the whole line; and when this is not the case, then the color representing the prevalence of shocks of the second order of intensity is given. The great circle of fire about Borneo as a centre, extending from Manila around to Sumatra, exhibits a broad belt of the deepest tint. The same is true of the region connecting the Andes with the Lesser Antilles. Iceland, the Azores, the Canaries, the Cape Verdes, parts of Italy, the country between the Mediterranean and the Caspian, — these are all regions of great earthquakes, and also, as we well know, of great volcanic eruptions. Of regions liable to great earthquakes and not volcanic, the following may be cited as the principal: portions of the coast of China, the region about the mouth of the Ganges, and that south of the mouth of the Indus, the Pyrenees, and the coast of Portugal between Lisbon and Oporto. If we should (as we very properly might) distinguish in the region of great earthquakes two divisions, — one in which highly destructive shocks may be expected to occur frequently, and the other where they take place only at long intervals, — we should then find that the former, that is to say, the pre-eminent earthquake areas of the world, are strictly limited to regions of volcanic activity, or to parts of the earth where such activity has only died out in the most recent geological periods. To say also that these regions of great earthquake shocks are almost exclusively in the neighborhood of the ocean is, then, almost unnecessary; since we know very well that there are almost no active volcanoes in existence except near the sea, — those reported in the Chinese annals as occurring in the Thian-Shan range, north of the Gobi Desert, being the only exceptions, while with regard to these there is much

uncertainty. The two most prominent facts, then, in respect
to regions liable to great earthquakes are, that they are almost
entirely coincident with areas of active volcanoes, and that
they also lie near the borders of the ocean.

Taking next into consideration the areas of earthquake shocks
of moderate intensity, we find that they also (since the greater
includes the less) are near coast-lines and volcanic centres,
either those now active or else such as have become recently
extinct, and especially that broad bands of the tint peculiar to
this class are found along many of the great ranges of mountains which are not volcanic, notably at the base of the Himalayas, the Alps, and the Pyrenees,— also on the northern edge
of the plateau of High Asia, through the belt of islands off the
east coast of Australia, from New Guinea around to New Zealand, and in the extinct volcanic islands of the Atlantic Ocean,
— as Ascension, St. Helena, Tristan d'Acunha, etc.

Here, before going any farther, it will be well to speak of the
very inaccurate ideas apt to be given by earthquake catalogues
of the real number and severity of shocks, and from a very
natural cause. In the regions where earthquakes are of rare
occurrence, and never severe, the slightest vibration is a matter of great interest, much talked about, and of course greatly
exaggerated ; especially are the newspaper accounts likely to
represent such uncommon events in the liveliest colors. In a
really earthquake-shaken region, like the west coast of South
America, and where at the same time newspapers are almost
unknown, by far the greater number of shocks are never put
on record, or at least they have not been until since the establishment of public scientific observatories, and of these
there are but very few, — while in thickly settled and highly
civilized regions each slightest jar of the ground has been
recorded. This leads to curious results in the records, and
might cause very erroneous conclusions to be drawn in regard
to the real earthquake character of different regions. Thus
it appears, *by the catalogues*, that nearly as many earthquakes
have taken place in Great Britain during the nineteenth century as in Chile, — while we know that in the first-named country no very destructive shock has ever taken place, and that
even minor ones are very rare. In Chile, on the contrary,

several very frightful earthquakes have occurred during the present century; one hundred and twenty-seven shocks were felt at Santiago in thirty-five months, and prudent people decline ever to sleep in a room with the door shut, lest it may become jammed by an earthquake, and egress be rendered impossible. It is a sufficient proof that a region is comparatively safe from real danger, to read as follows: "Newport, R. I., 1766, August 25. Violent shock. *No damage done.*" A violent shock not producing any damage would be a desideratum on the South American coast, where probably the record would have stood: "*A very slight vibration. No damage done.*"

It is from considerations like those just suggested, and also, in some degree, through the absence of reliable data, and the habit which our newspapers have of exaggerating all events, whether physical or political, that we may account for the seismological character of our own country being very much misrepresented on Mr. Mallet's map. On it we find the whole region east of the meridian of 95°, excepting a small area about the Upper Great Lakes, colored of the middle tint, indicating a region of considerable earthquake activity. This band is also extended up to the head of the Missouri River, over a belt of country two hundred miles wide. On the Pacific coast, on the other hand, a broad band of the lightest tint, indicating only an occasional visitation of the lightest possible shocks, extends back from the ocean as far as Salt Lake, including the region north and south between the mouth of the Colorado and the northern line of California. This tint is also continued up the coast to the Aleutian Islands, excepting only a small area about the mouth of the Columbia River, where two active volcanoes are inserted, and, apparently in consequence of their presence, the region thereabout is colored of the middle tint.

Now certainly our personal experience — and this is well supported by the catalogues — shows that we live in a region where earthquakes are very little to be apprehended, and where there is no record of any destructive one ever having taken place. We have no remembrance of ever experiencing a shock, even of the slightest kind, in Massachusetts.

We have to go back as far as 1755 to find any record of a decided earthquake, and this seems to have been connected with a great agitation extending over a large part of the earth's surface, and it is not unlikely that the focus of disturbance was far out at sea. It will be remembered that the " Great Lisbon Earthquake " commenced November 1, 1755, and that it was one of the most violent and widely extended on record. The shocks continued, at various places around the Mediterranean, with occasional intervals, for many months, and nearly the whole circumference of the Atlantic Ocean was in a disturbed condition, while portions of the East India Islands were also vibrating synchronously, if not sympathetically, with the other side of the globe.

The shock of November 18, 1755, was felt all along the Atlantic coast, between Halifax and Maryland, and west certainly as far as Lake George, in New York. It was quite severe at Boston, — more so, probably, than anywhere else within our territory. This earthquake was described by Professor Winthrop, of Harvard College, in the Philosophical Transactions of the Royal Society (Vol. L. Part I. p. 1), with considerable detail, and with no little skill and critical acumen. The shock was violent enough to throw down a considerable number of chimneys and the gable ends of some brick buildings. Throughout the State many stone fences were more or less injured. Some cracks were made in the ground near Scituate, from which water and sand issued, to the extent of " many cart-loads." Previously to that, in 1638, 1658, 1662, and 1727, shocks had been felt in Boston, of which that of 1662 was severe enough to throw down some chimneys. It does not appear, however, that any person has ever been killed by any earthquake in New England ; so that it is pretty safe to conclude that some, if not all, of the chimneys thrown down had been built with very poor mortar. We are not aware that there is any record of any considerable shock having taken place in New England since that of 1755.

On the Pacific coast of the United States, however, hardly a year elapses without some pretty severe shock. The number of earthquakes recorded in California for the thirteen years ending December, 1863, is one hundred and eleven. Many very heavy ones have occurred there since the beginning of the present century. In 1812 the whole southern part of that State was violently agitated during four and a half months. In some regions the inhabitants abandoned their houses altogether. Several of the " Missions " — substantial stone buildings — were thrown down; in one — that of San Luis Capistrano — religious services were going on at the time, and many of those present were killed, the number of persons thus perishing being stated by various authorities at from thirty to forty-five. A number of lives were also lost at the Missions of La Purissima, one hundred and twenty miles distant from that of San Luis Capistrano. It must be recollected that the State, then the Mexican province of Upper California, was at that time extremely thinly inhabited. Had it been a populous region, it would seem, from the descriptions of the character of the shocks, that the loss of life and property must have been very great. Among the earthquakes which have happened in California since it became a part of our own territory, two are particularly to be remembered, — those of October 8, 1865, and of October 21, 1868. The first did considerable damage to property in San Francisco, and the other was severely felt over an extensive area, demolishing a great number of buildings in that city, and especially in the towns on the opposite side of the bay between Oakland and San José. Several persons were killed by falling fragments. In view of the above-cited facts, it will readily be seen that even coloring both sides of the United States as equally liable to seismic demonstrations would not at all be supported by the facts; while representing the Atlantic coast as more shaky than the Pacific slope is very far out of the way.

The extending of a band of color indicating serious earthquake action up the Valley of the Missouri is also quite unsupported by the facts. It is true, however, that a region of a few hundred square miles in area, near the junction of the Mississippi and the Missouri Rivers, was subjected to violent earthquake shocks during several months, in the years 1811–12, — a remarkably exceptional case in every respect, and therefore worthy of a brief notice. This disturbance commenced December 16, 1811, with an earthquake which was felt over a large portion of the Valleys of the Mississippi, the Ohio, and the Arkansas; it was also noticed as far to the southeast as Florida, although the shocks were feeble to the east of the Alleghanies. New Madrid, on the Mississippi, in latitude 37° 45', a little below the mouth of the Ohio, seemed to be the focus of the disturbance, and the shocks continued there daily and almost hourly for several months; they are reported as having finally ceased about the time of the great earthquake of Caracas, March 26, 1812. As a result of this, a large tract of country west of New Madrid, extending seventy or eighty miles north and south, and thirty east and west, was permanently sunk considerably below its former level, and converted into a marsh. This was truly an interesting and peculiar occurrence, as it is almost the only instance on record of a region far from volcanoes, from mountain chains, and from the ocean, being subjected to a long and violent disturbance of this kind; it is also remarkable that the heavy shocks in this locality have been repeated only once, so far as we are able to learn,—namely, in 1865, August 17, when a considerable part of the Mississippi Valley was shaken with some violence, although no serious damage was done at any point. The shock was most distinctly felt at New Madrid. It is said that light vibrations have frequently occurred in the district of the "Sunk Country," as it is called, since the great ones of 1811-12.

An examination of the catalogues and maps of earthquake areas, with a view to their correlation with the geological structure of those areas, shows some very interesting facts. It is clear that persons living on the older geological formations have much less reason to apprehend earthquake disturbances than those who have under them the more recent members of the series. There is hardly any region liable to severe shocks where there are not, in the vicinity at least, large accumulations of strata belonging to one of the later geological epochs. Chains of mountains made up of Palæozoic rocks, as, for instance, the Urals, the Appalachians, the Brazilian ranges, the Scandinavian mountains, and the Laurentian mountains, are never the scene of violent or destructive shocks.

Where the newer formations do exist, but where, however,

they remain undisturbed, or nearly in the same relative posi
tion in which they were deposited, there, too, is immunity from
earthquake damage, and, *a fortiori*, where the older formations
occur and are entirely undisturbed.   Thus, the whole vast region of the Central and Northern-Central portions of North
America, north of the parallel 40°, is remarkably free from
earthquakes, and we have there one of the largest areas in the
world of nearly horizontally stratified Palæozoic rocks.   From
the western base of the Appalachian chain, towards the northwest, over a wide belt, including the Upper Great Lakes, and
trending off towards the Arctic Ocean, there extends a tract
embracing many hundred thousand square miles, and included
between the eastern base of the Cordilleras and Hudson's Bay,
over which only the oldest geological formations occur, and
where these have remained almost wholly undisturbed since
their original deposition.   This is a region left entirely uncolored in the seismological maps, and which, so far as can be
learned, is indeed almost, if not quite, exempt from even the
minor earthquake shocks.   So, too, the region of the Plains
in our own territory, although underlaid by the more recent
formations, is little, if at all, troubled by earthquakes, and we
know that the formations are here also almost horizontal.   The
same conditions may be traced all over the world, so far as our
information goes ; so that we are justified in asserting that it
is extremely rare to find earthquakes occurring over geologically undisturbed areas, or regions where the strata have not
been turned up and folded, and that the same is true even
where such geological disturbances have taken place, provided
they have not been continued down to a recent geological
period.

   Thus we have shown, that, from a geographical point of view,
great earthquakes, and even those of minor consequence, are
clearly connected in their place of occurrence with the position
of the oceanic basins, with the existence of great mountain
chains, and consequently with the distribution of volcanoes ;
also, that they are unmistakably associated with the existence
of the more recent geological formations, and with their most
disturbed condition.   Hence we have the strongest reasons for
believing that earthquake phenomena are dependent on gen-

eral laws, such laws as have governed the building up of con-⟩
tinents and the bringing of the great features of the earth into
their present stage of development. They cannot be mere local phenomena, occurring without any mutual relations to each
other, or as disconnected with the whole series of geological
events which scientific investigations show so clearly to have
been governed by a law of progress. The same conclusion
may be drawn from a consideration of the extent over which
many earthquake shocks are felt, — the magnitude of the
area shaken, and its proportion to the whole surface of the
earth, being considered, very fairly as it would seem, a decided indication of the magnitude of the cause. One of the
greatest earthquake shocks on record is that already referred
to as "the Great Lisbon Earthquake," the centre of disturbance
having been situated near the coast of Portugal, and the effects
of the shock having been most fearful at that city. This earthquake produced sensible effects over an area of the earth's surface included between Morocco on the south and Iceland on the
north, Töplitz in Bohemia on the east, and the West India Islands on the west. The great earthquake of August 13, 1868,
of which, however, only the most unsatisfactory accounts have
yet reached us, appears to have been felt along the Andes, over
a breadth of forty degrees of latitude, and its effects were distinctly visible in the great waves it raised at Juan Fernandez,
on all the Hawaiian Islands, on the coast of Japan, and even in
Australia and New Zealand.

On examining the phenomena of earthquakes with reference to the time of their occurrence, various interesting results
have been obtained, as respects their frequency, both at different seasons of the year, and while the earth is in certain
positions with regard to the sun and moon. It appears, also,
that there are certain periods during which the earth is in a
peculiarly disturbed condition, and that not unfrequently a large
number of shocks take place at about the same time in regions
far removed from each other. As a marked instance of this,
may be mentioned the latter part of November, 1852, when a
large portion of the Pacific coast, both of North and South
America, was in motion, at the same time with the whole of
the East Indian Archipelago and various intermediate places.

This earthquake period commenced in the East Indies, in Southern Sumatra, on the 11th of November, and the shocks continued in various parts of the Archipelago until the 26th, when the great one took place which was felt all over the East India Islands, from Manila to Sumatra. The disturbance was kept up through the whole of December, and, on the 21st of that month, had, in the island of Java, reached a degree of violence exceeding anything previously known. From the 27th to the 30th of November the earth was in constant motion in all the East India Islands. During exactly these days, — that is, from the 26th to the 29th of November, — tremendous shocks were constantly felt in the Great Antilles. On the 26th of November, very severe earthquakes agitated the Pacific coast of North America, from Mexico to Northern California, and indeed the whole region between the Colorado River and the coast was in a state of continual vibration for nearly two months. On the same day, November 26th, an earthquake was felt in Italy; the next day, a slight shock on the Atlantic coast of the United States, and a heavy one on the South American coast; and still the next, another in Chile. It would appear that at this time both sides of the Pacific Ocean, from China to Australia on the west, and between California and Chile on the east, were vibrating synchronously and extensively, and that this condition of things lasted for nearly two months, while several points in other regions were also seriously implicated in the disturbance. This was undoubtedly one of the grandest epochs of earthquake disturbance which have ever been known, and it is hardly possible to explain the synchronous occurrence of so long-continued and violent a series of shocks in the regions affected by simply considering it as an accidental coincidence. A great many other instances might be cited of earthquake disturbances taking place at the same time, in regions far distant from each other; while, on the other hand, it is true that severe shocks have often taken place which were limited to quite a narrow area.

The coincidences of earthquakes and volcanic activity are curious, and not easily brought into harmony with any theory. The great fact is clear enough, that by far the most severe and the most frequent earthquake shocks are in countries of volcanic activity. But it is also not to be denied that volcanic eruptions do occur occasionally in perfect quiet, so far as vibrations of the adjacent crust are concerned. The same uncertainty exists with regard to the internal connections and sympathy of volcanic vents, whether at a distance from or near to each other. Cases have repeatedly occurred where adjacent volcanoes have not sympathized in the slightest degree in their periods of rest and activity, even when in immediate proximity to each other. One of the most curious of these instances is that of the summit crater of Mauna Loa, and Kilauea, the famous side-crater on the same mountain, nearly ten thousand feet lower down. It has repeatedly happened that the upper one has been in violent eruption, while the lower was in no degree more active than usual, thus showing that the two great vents of the same volcano were not in immediate connection. On the other hand, it has often occurred, that, of two volcanoes near each other, or even at a considerable distance apart, one has become absolutely quiet at the very moment when the other has suddenly burst into eruption ; the instances of this kind are, some of them, so marked, and the correspondences in the commencement or termination of the seasons of activity have been so exact, that it would be quite impossible to pass them over as mere accidental coincidences.

The question has been much discussed, whether volcanoes in reference to earthquakes act as " safety-valves," — that is, whether their eruptions, once commenced, can be looked upon as in any degree removing the probability of violent shocks. That such is the case is the almost universal belief through earthquake-shaken regions in the neighborhood of great volcanoes. Indeed, it seems not unreasonable to suppose, that, the internal forces having once found vent for their energy in the eruptive action, the vibration of the crust, which can hardly be looked on as anything else than the result of the struggles between expansive force on the one hand and the pressure and tenacity of the superincumbent material on the other, would be suspended. On examining the records, it will be found that there are many instances which show clearly that earthquake shocks, previously severe, have ceased entirely at the moment of the eruption of some adjacent volcano ; while there are other instances in which severe earthquakes have been felt some time after great eruptions in the vicinity had commenced. In the very numerous instances where volcanic eruptions have been the signal for the stoppage of a long series of earthquake shocks in the vicinity, it is difficult to admit any other explanation than that the issuing of the lava has relieved the pressure and thus removed the cause of the shocks ; while in the cases of an opposite character, where the vibration still continued after the eruption had begun, it is reasonable to suppose that the relief was only local, and not sufficient to affect the whole adjacent region.

As before remarked, the curves indicating the number of recorded earthquakes in all parts of the world expand rapidly as we approach the present epoch. There is no reason to suppose, however, that this means anything more than that our records are growing every year more complete : only the observations of the last century and a half can be considered as making the slightest claim to completeness. No inference can be drawn at present, then, or probably for a long time to come, as to whether seismic energy, as a whole, throughout the world, is on the increase or decrease. On this point we shall be for a long time in the dark. But the question next arises, whether the records, especially those of the last two or three centuries, exhibit, when plotted in curves, any indication of irregular or paroxysmal energy ; that is, whether there are certain epochs during each century, when the number and intensity of shocks are greater than at others. Although the dates are far too incomplete to admit of a perfectly satisfactory answer to this question, Mr. Mallet thinks he is justified in asserting that there are minor intervals of comparative repose, averaging from five to ten years in duration, alternating with periods of considerably increased activity of a year or two in length. These shorter intervals do not seem to follow any regular law, so far as can be made out from the curves ; but they seem to be in connection with periods of fewer earthquakes, and usually with the occurrence of less violent shocks. There are also two very well marked epochs of extreme violence and frequency of earthquakes, — one towards the end, and one, still more violent than the other, about the middle of each century. The form of the curves seems to indicate a comparatively sudden burst of seismic energy at each great paroxysm, and then a more gradual subsidence of the action ; as if the disturbing forces had been of a nature to reach rapidly the maximum of their power, and then to sink more slowly into their normal condition of activity. Still, the data are few for general results of much weight in. regard to long periods of alternate repose and paroxysmal energy. If, as Mr. Mallet thinks, such conclusions can already be drawn, it is a strong argument in favor of considering earthquake action to be connected with some great general cause, commensurate, in the magnitude of the area in which it acts, with that of the earth itself.

In comparing the relation of earthquakes to the times of the year in which shocks have occurred, in order to ascertain whether there are months or seasons in which seismic energy is more developed than in others, quite interesting coincident results have been obtained by all who have occupied themselves with these investigations. In the first place, it is clearly made-out that there are more earthquakes, in the northern hemisphere, during winter than summer. Thus, Dr. Kluge gives, for the shocks registered from 1850 to 1857, nineteen hundred and eighty-four as occurring in the winter half of the year (September to March), and only eighteen hundred and thirty-four as taking place in the summer half. The months in which the smallest number of shocks took place were May, June, and July ; and October, December, and February, those in which the number was greatest. Mr. Mallet draws substantially the same results from the comparison of the curves of mensual seismic energy for the whole period of the catalogue, or thirty-two centuries. In the northern hemisphere he finds the annual paroxysmal minimum to occur in July, and the maximum in January, while the preponderance of winter over summer in the number of shocks is very decided. The results of observation in the southern hemisphere agree with those in the northern, the frequency of earthquakes there being greater in summer (our winter) than in winter (our summer) ; but the observations are so limited in number, and the area is so much more extensively covered by water, that at present any deductions of this kind in regard to the southern hemisphere have much less weight than the similar ones for the northern. The same results are shown when the months are grouped into four seasons, — the curves showing clearly a maximum for the three winter months, and a minimum for the summer.

Another coincidence appears to have been pretty clearly indicated, if not positively made out, by the labors of Messrs. Mallet and Perrey : namely, the occurrence of a maximum of earthquake shocks about the time of the winter solstice, and a minimum at the autumnal equinox. And there is still another branch of inquiry with reference to the frequency and violence of earthquakes, which is of great interest, although as yet by no means thoroughly worked out : that is, the action of the moon on the earth, or the connection between the phases of the moon and the recurrence of shocks. The coincidence of certain great earthquakes with extreme high or low tide had been repeatedly noticed in South America many years ago, and the probable influence of the moon on the interior of the earth asserted by different scientific authorities. Baglivi, an Italian author, in his description of the Roman earthquake of 1703, published in 1737, notices particularly the fact of the more common occurrence of earthquakes at the time of full moon. M. Perrey was the first, however, to enter into the laborious calculations necessary to throw further light on this question of so much interest ; and although it cannot be considered as thoroughly settled, still the facts seem to indicate that the action of the moon, or of the sun and moon combined, is really perceptible in increasing the number and violence of earthquakes at certain periods. M. Perrey's results, as obtained from a combination of the observations of 1844 – 47, are as follows : *First*, that earthquakes occur more frequently at the Syzygies than at the Quadratures ; *secondly*, that they also are more numerous at the Perigees than at the Apogees ; and, *finally*, that, whenever a disturbance is going on, the frequency of the shocks is increased by the passage of the moon over the meridian of the place in question. These results would indicate that the moon has an action on the interior of the earth somewhat analogous to that which it exerts on the ocean, — the time of greatest frequency of shocks agreeing with that of the highest and lowest tides. The great interest of this investigation will be easily understood, since it bears very directly on one of the most vexed questions of modern geological science, namely, whether the interior of the earth is really in a liquid state, or sufficiently so to admit of its yielding to the attraction of the sun and moon in such a degree as to produce a sensible result, as would be the case, provided it could be clearly proved that the supposed lunar influence on the frequency of earthquakes really existed. Such an investigation, moreover, has an important bearing on many points of theoretical geology, and it will certainly not be dropped until the question has been definitely settled. Of M. Perrey's conclusions Mr. Mallet says, that they rest upon so narrow a basis of induction that they must be accepted with caution ; yet he admits that they possess more than sufficient probability, from physical considerations, to induce further inquiry. The Committee of the French Academy of Sciences to which M. Perrey's conclusions were referred were evidently much impressed with the character of his results, although cautious in accepting them, until they should be confirmed by the reduction of future observations, or by going back and computing a still greater number of older ones.

However important the relative frequency of earthquakes, as compared with the positions of the sun and moon, may be to the scientific man, as having a profound theoretical significance, people generally are much more interested in the connection of seismic with meteorological phenomena. A great many persons think that they remember some peculiarities of the weather as having preceded any great shock ; and in almost every earthquake-shaken region there are popular theories as to the premonitory symptoms of these disturbances, — although these are very different in different places. The most common one is, perhaps, that oppressive heat, accompanied by unusual stillness of the atmosphere and a light mist, is a sure forerunner of a shock. In accordance with this theory, the inhabitants of San Francisco were greatly excited, last September, by the occurrence of a remarkably smoky appearance in the atmosphere during several days ; and a report having been set afloat that an uncommonly high tidal wave had been experienced in the harbor, the city became wild with excitement. Nothing unusual happened, however, and the smoke was afterwards traced to burning forests far north on the coast. The most careful comparison of the catalogues of earthquake occurrences with registers of the weather has failed to reveal any substantial reason for supposing that any of these peculiar indications really do precede the shocks. Only this much appears probable: that a great depression of the barometer, implying a diminution of the pressure of the atmosphere on the earth, may be in some cases the determining cause of an earthquake. This, as we can easily conceive, might be the case; since, if we suppose the normal condition of the crust of the earth in an habitually disturbed region to be that of a nicely balanced equilibrium between the internal forces seeking exit, or relief by change of place, and the pressure of the overlying material, gravity and tenacity acting against expansion, it is not unreasonable to admit that a sudden depression of the barometer, perhaps to the amount of two and a half inches, equal to one twelfth the whole weight of the atmosphere, may turn the scale, so that the crust shall give way and the pent-up forces find relief, giving us the evidence of it in a vibration of the superincumbent strata. There are many facts which seem to indicate that the severe storms, gales of wind, and heavy rains, which have repeatedly been observed to occur simultaneously with earthquake shocks, and which, from meteorological causes, are preceded by a remarkable fall of the barometer, are thus causally connected with seismic disturbances. The depression of the mercurial column indicates a change in the currents of the atmosphere, which will result in a violent storm, and the diminished pressure of the atmosphere is the direct agent in starting the vibration, which takes place sooner than it would have happened, had it not been for this disturbing element.

Many curious statements have been made with regard to the presentiments of approaching earthquakes manifested by different animals, some of which seem well authenticated, while others must be set down as the results of excited imaginations. Some of the peculiar actions ascribed to animals may easily be accounted for by the emission of carbonic acid or other gases

from the ground, which is known to accompany some earthquakes in volcanic regions, and which might be perceptible to
animals whose sense of smell or nervous susceptibility was
more delicate than our own. Dogs are supposed to be peculiarly sensitive in this respect, and hogs and geese are believed
to show fear of approaching volcanic disturbances sooner than
other animals. Birds generally are very quick at taking alarm,
as might naturally be expected from their delicate organiza-
tion. All incidents recorded with regard to the behavior of
animals, before and during earthquake shocks, must be taken
with many grains of allowance ; but such as are well authenticated are extremely interesting, as indicating differences between the nervous susceptibilities of man and the lower animals.

Whether there is any relation between earthquake phenomena and the magnetism of the earth is a question which has
been frequently discussed, and for the satisfactory answering
of which the data are not yet sufficient. We know no reason
why there should be any real connection between the disturbances of the earth's crust and the magnetic currents which
circle around it, nor has any been proved. On the contrary,
most, if not all, of the investigators in this branch consider
that there is no reason to believe that the unusual vibrations
of the magnetic instruments, which have been sometimes observed in earthquakes, are anything more than the mechanical result of the motions of the earth's crust.

We have now gone rapidly over most of the ground which
has been occupied by compilers of earthquake catalogues, and
given a sketch of the principal results. It must be remembered, however, that a large portion of the data used are entirely wanting in the elements of scientific accuracy, and that
in consequence of this looseness of statement only conclusions
of the most general character could be drawn from them. So
impressed was Mr. Mallet with this fact, that he thus expressed
himself at the end of the Report to the British Association
which had occupied him for so many years. He says : " In
conclusion, I would repeat my conviction that a farther expen-
diture of labor in earthquake catalogues, of the character hitherto compiled, and alone possible from the data to have been

compiled, is now a waste of scientific time and labor. The
main work presented for seismologists in the immediate
future must consist in good observations, with seismometers
advantageously placed at sufficiently distant stations, and galvanically connected as to time,—and in the careful observation
of the traces left by great shocks (when of recent occurrence)
upon buildings, and other objects, artificial and natural, with a
view to determining the nature of the forces that have affected
them; aided by the resources of the physicist and the mathe-
matician."

Just about the time the above-quoted conclusions of Mr.
Mallet were put upon paper, there occurred the great earthquake of December, 1857, which shook a large part of the
Neapolitan territory, and was the third in extent and severity
of all those of which there is any record as having occurred in
Europe, — since more than ten thousand persons were killed
by it, and a great number of towns and villages were almost
destroyed. Immediately after this calamity, Mr. Mallet applied
to the Royal Society of London for a small grant of money, to
pay a part of the expense of visiting the locality, and making a
thorough investigation of all the facts in the light of the most recent seismological inquiries. The request was acceded to, and
Mr. Mallet travelled carefully over the shaken region during
several months, and was afterwards employed for nearly two
years in preparing his report, the title of which stands at the
head of this article. This report was published in 1862, the
Royal Society contributing three hundred pounds towards the
expense. It fills two royal octavo volumes, and is most elaborately and beautifully illustrated, in a manner worthy of the
first really thorough investigation in the department of Seismology.

It is hardly necessary to state that one investigation has not
exhausted the subject; it has rather set the example of what
ought to be done for many earthquakes ; and it is especially of
value, as leading the way in a new line of research, and as
showing what can and must be done in order to arrive at as
complete a knowledge as possible of the workings of the mysterious agencies by which these great convulsions are brought
about. Some of the more important results obtained by Mr.
Mallet in regard to the Neapolitan earthquake may here be

given, as a specimen of the kind of material which will have to be accumulated from all quarters of the globe before the demands of scientific accuracy shall have been satisfied.

In the first place, in the map accompanying the report in question, the regions in which the shock was equally intense are designated by curves, called *isoseismal curves ;* then the whole of the wave-paths, or lines of direction in which the shocks were propagated at each locality, are marked by red lines. These wave-paths of course radiate from the focal point of the shock, and so carefully were they determined, chiefly by observations of the position of fallen buildings, and the character of the movements and fractures in those left standing, that sixteen of these lines, when protracted back, pass through the same focal point, or within a circle of five hundred yards radius around it, while thirty-two more fall within a circle concentric with the former and of one mile radius. Now, theoretically, the intersection of any two wave-paths is sufficient to fix the position of the " seismic vertical," or the point on the earth's surface vertically above the spot where the impulse or shock originated. The evidence, then, in this case was ample for determining this point as accurately as possible; since, whatever be the nature of the impulsive force, or however it may operate, the wave of impulse, as propagated outwardly, passes simultaneously, or almost so, from points about the actual focus at a considerable distance from each other, — the point from which the disturbance starts not being, by any means, a mathematical one. The position of the point on the surface vertically over the seismic focus was found, as above, to be near Caggiano, a village sixty miles a little south of east from Naples.

The next important question to be settled was the depth of this focus below the surface, — a point of great interest, as will be perceived at once, in its connection with the theory of earthquake action. This depth can easily be obtained by mathematical calculation, when the distance on the surface from any station to the seismic vertical is known, together with the angle of emergence of the wave-path, the seismic vertical being another wave-path, and the point of convergence of the two being the focus from which the wave started. Of course the limits of error are considerable in an investigation of this kind ;

but the results, as graphically exhibited on Mr. Mallet's diagram, are quite as satisfactory in their agreement as could be expected. Out of twenty-six separate wave-paths, twenty-three start from the seismic vertical at a depth of above 7$\frac{1}{4}$ miles ; the maximum depth is 8$\frac{1}{8}$ miles, and the minimum 2$\frac{3}{4}$ miles.* Eighteen of the wave-paths start from the seismic vertical within a vertical range of twelve thousand feet, and having a mean focal depth of 5$\frac{3}{4}$ miles, which may be taken as the depth of the focus. Here is an extremely important numerical result, and similar results from other regions are highly desirable for comparison with this.

It will be impossible here to enter into the detail of the other numerical results obtained by Mr. Mallet, — the position and depth of the focal centre being, of course, the most important, and having been determined in this instance for the first time with any approach to accuracy. Other interesting points discussed in the summing up of the results of the investigation are : the form of the isoseismal areas, — that is, of the regions over which the shock was felt with equal intensity ; the relations of this area to the focal depth ; the effects of the physical configuration of the surface and the geological structure of the region on the progress of the wave ; the proofs of reflection and refraction of the shock by a range of mountains standing in the way, including reasons why certain areas escaped entirely ; the form, position, and dimensions of the focal cavity ; the amplitude and velocity of the wave, both on the surface and in the wave-paths ; the velocity with which the shock started, and its gradual dying out ; the relation of the seismic foci of the Italian Peninsula, and the general relations of the seismic bands of the Mediterranean basin. To give even a synopsis of the results obtained under the above heads will not be possible here ; those who desire to investigate seismic phenomena must consult the volumes themselves.

We see that Mr. Mallet was fully justified in demanding more thoroughly scientific observations than those we had previously to his work, and that he has given a most excellent example of how such investigations should be made. He has shown that we can not only learn much from the application

---

* These results are given in geographical miles, of sixty to a degree.

of seismological inquiries to future earthquake shocks, but that we have it in our power, to a certain degree, to recover the history of the past, by investigating the results of former convulsions as registered in the buildings fissured or in the ruins of those overthrown by ancient earthquakes.

Among the practical results of investigations like those of Mr. Mallet, there are none so interesting to the public at large, especially to persons living in earthquake regions, as those which relate to the proper methods of structure for safe houses and other edifices in countries liable to these disturbances.

This experienced observer expresses his strong conviction, " that the evils of the earthquake, like all others incident to man's estate, may be diminished, or even multiplied, by the exercise of his informed faculties and energies, and by his application of forethought and knowledge to subjugate this, as every other apparent evil of his estate, by skill and labor." He further adds, in reference to this important question : " Were understanding and skill applied to the future construction of houses and cities in Southern Italy, few, if any, human lives need ever again be lost by earthquakes, which there must recur, in their times and seasons."

What is true of Southern Italy should also be true of the Pacific coast of our own territory, a region liable to severe shocks, and yet where we hope to see populous States develop themselves in wealth, intelligence, and security to life. The prevailing tone in that region, at present, is that of assumed indifference to the dangers of earthquake calamities, — the author of a voluminous work on California, recently published in San Francisco, even going so far as to speak of earthquakes as " harmless disturbances." But earthquakes are not to be " bluffed off." They will come, and will do a great deal of damage. The question is, How far can science mitigate the attendant evils, and thus do something towards giving that feeling of security which is necessary for the full development of that part of the country ?

There has repeatedly been talk at San Francisco of establishing an astronomical observatory, either by itself or in connection with the State university. If the people of California are wise, and have money to give for scientific research, let them found a physical, and not an astronomical, observatory. We have enough of the latter already, ill-equipped, and in the majority of cases not manned at all. Quite a sufficient number of large telescopes are rusting on their piers in various parts of the country, as valueless for all real scientific results as if they never had been taken from the boxes in which they were imported. Let California take warning from these, and remember that a very large endowment is necessary for the permanent maintenance of an astronomical observatory, and that, if not permanently maintained, in the hands of an able astronomer, with the means of paying his assistants and of publishing his results, it will be nothing but an expensive toy. Besides, the climate of California and the climatological conditions are ill-suited to astronomical work in a fixed observatory. The fogs of San Francisco, and the dust of the interior, will be found alike unfavorable to the successful prosecution of this branch of scientific research. A physical observatory, on the contrary, which need not necessarily be a permanency, having as its principal object the investigation of the seismological phenomena occurring on the Pacific coast, would, if properly managed, furnish results of exceeding value, not only as contributions to an important branch of science hitherto much neglected, but as having a practical bearing on the welfare of the people and the development of the State, the value of which can hardly be overestimated. In no portion of the world is there a better chance for an establishment having in view the thorough investigation of earthquake phenomena. The great plain of the Sacramento and the San Joaquin should for a time be connected with San Francisco galvanically, by wires proceeding from the branch observatories at properly selected localities. Seismometers of the most approved construction should be set up, and their records compared with the other results of every important shock, as shown in the effect on buildings and on the surface of the ground, and in all the other methods of which Mr. Mallet's book furnishes so excellent a model.

Of Herr Volger's volume and theory something may be said at another time, in discussing the various theories of the nature of the forces involved in the phenomena of volcanoes and earthquakes.

VOLCANOES.

1. *Vesuvius.* By JOHN PHILLIPS, M. A. Oxford. 1869. 12mo.
2. *Histoire Complète de la grande Éruption de Vésuve de* 1631. Par H. LE HON. Bruxelles. 1866. 8vo.
3. *Reise der Oesterreichischen Fregatte Novara um die Erde in den Jahren* 1857, 1858, 1859. Geologischer Theil, Erster Band, Erste Abtheilung, Geologie von Neu-Seeland. Von DR. FERDINAND VON HOCHSTETTER. Wien. 1864. 4to.
4. *Voyage Géologique dans les Républiques de Guatemala et San Salvador.* Par MM. A. DOLLFUS et E. DE MONT-SERRAT. Paris. 1868. 4to.
5. *The Natural System of the Volcanic Rocks.* By BARON F. RICHTHOFEN. Extracted from the Memoirs of the California Academy of Sciences. San Francisco. 1868. Pamphlet.

WE have placed at the head of this article the titles of a few of the many volumes devoted chiefly to the subject of volcanoes which have issued from the press during the past few years. To give a complete list of the volumes and papers in which the phenomena of volcanism have been described and discussed, even if only the productions of the last five years were to be included in it, would require many pages. On the subject of the volcanic island of Santorin alone, at least six different works were published during the year 1868. One author, Le Hon, gave, in 1866, a complete history of an eruption of Vesuvius which took place two hundred and thirty-eight years ago; while several other writers, some of them known as geological authors and others not, have taken advantage of the recent period of activity of that interesting volcano to serve up portions of the mass of the old material in a new form, adding in some cases new facts of value to the previously existing stock, but generally relying for their chances of success rather on elegance of typography, or other extrinsic circumstances, than on scientific accuracy or originality of ideas. The reason of the exceptional activity in this department of book-making is partly that the volcanoes themselves — at least several of those best known — have been unusually active, and partly because the fashion of illustrated and sensational books on scientific subjects has been set, and of all the subjects which geology presents there is none which so excites the popular mind as the phenomena of volcanoes and earthquakes.

Earthquakes are events simply fearful; there is nothing about them which is not appalling in its nature. They come without warning, and leave nothing but dismay and ruin behind. Even the minor shocks are terrible, and more alarming in proportion to the number of times they have been experienced. It is only in California that an attempt has been made to pooh-pooh an earthquake; but even there the hollowness of the derision was but too evident. In an earthquake-shaken country the time that elapses between the instant when one perceives that an earthquake wave is approaching and that when its first effect is felt is one into which a thousand apprehensions can be crowded. Then, if ever, one feels the utter insignificance of man as an integral part of creation. The blow may fall lightly and leave no sensible trace behind; or, on the other hand, it may crush and overwhelm. The regulating screws of the horrid machinery are invisible. There is no reason why one should await with more calmness the approach of an earthquake shock than, with his head on the anvil, the falling of a steam-hammer, not knowing beforehand at what point the ponderous mass is to be arrested by the engineer in charge of the machine.

Volcanoes, on the other hand, give in almost all cases some previous warnings of their intention to change their usual quiescent state for one of destructive activity. Their disastrous effects can often be to a large extent avoided by flight. It is only very rarely that an eruption is so sudden and violent as to overwhelm and destroy without previous and oft-repeated warnings. Again, eruptive volcanic action is usually prolonged over many days or weeks, or even months, and the phenomena exhibited are usually — if the eruption is on a large scale — of surpassing grandeur, from a picturesque as well as from a scientific point of view. Perhaps there is no scene offered by any play of nature's forces so wonderfully attractive as that of a great volcanic eruption, especially when seen by night. The combination of every conceivable element of the picturesque and the sublime afforded by the great outbreaks of Kilauea, as reported by the few who have had the good luck to witness some of them, may be mentioned as an instance in point.

No wonder, then, that the subject of volcanoes has always been an attractive one to the general, as well as to the scientific, traveller and writer, and that such a great number of volumes have been published, and are still publishing, treating either of volcanoes in general or of particular eruptions or periods of eruptive activity. The work of the veteran Oxford professor, John Phillips, the title of which is placed at the head of the list preceding this article, is one of the most noticeable of those possessing a somewhat popular character. Within the limits of three hundred and fifty pages it gives a succinct history of Vesuvius and of the adjacent volcanic region so much visited by travellers, and is on all points exact and clear. The illustrations of the volume are numerous and effective, although not elaborate, and very far from sensational. The book is exactly what was desirable as a guide to travellers of scientific tastes, and may be consulted with profit and pleasure by the professional geologist. It contains, besides, a catalogue of Vesuvian minerals. There is also a chapter devoted to the theory of " volcanic excitement," — a subject on which much has been written, especially of late, but in regard to which it must be admitted that we have still much to learn.

The work of M. Le Hon, placed second on our list, is especially valuable as containing a large map, which appears to have· been carefully constructed, and which exhibits all the flows of lava from Vesuvius between the years 1631 and 1861. This is the only map which professes to give with any approach to exactness the position of these masses, and evidently it could not have been produced without considerable labor and

without numerous excavations. The description of the eruption of 1631 is carefully compiled, and gives a good idea of
this the most devastating of all the modern outbreaks of Vesuvius. By this eruption it is probable that at least four thousand persons lost their lives in various ways, while more
than forty towns and villages were destroyed, the pecuniary
losses being estimated at twenty millions of ducats, — an enormous sum at that time.

The volcanic phenomena of a far distant but exceedingly interesting region — New Zealand — are brought to our notice
for the first time in a comprehensive manner by Dr. Hochstetter, in two separate works, — ohe, in royal 8vo form, of a popular character, entitled simply " Neu-Seeland " ; the other, a volume of the series published by the Austrian government as the
official account of the voyage of the frigate Novara, made in
the years 1857–59. The first-mentioned work was published by
Cotta, in 1863, with every luxury of adornment, and is one of the
most attractive books — half scientific and half narrative — ever
issued. The quarto official volume is also beautifully printed
and illustrated, and is largely devoted to a description of the
New Zealand volcanoes, as well as of the wonderful geysers, hot-
springs, and solfataras which form so peculiar and attractive a
feature of the island, and which are admirably represented in
the chromo-steel plates of the popular volume and the chromo-
lithographs of the other. These indicate a type of geological
scenery resembling that of the geysers of Iceland, but on a
grander scale, and with the peculiar added beauty of a wonderfully interesting and abundant vegetation. Dr. Hochstetter's books are rich in information about a new and remarkable
region, but they are very little encumbered with generalities
or theoretical views.

Almost equally magnificent in its typography and style of
publication is the work placed fourth on our list, — an official
publication of the French government, issued from the *Im-
primerie Impériale*, as an instalment of the results of the
scientific mission instituted by the Emperor for exploring
Mexico at the time when his unfortunate military expedition
to that country was planned. In carrying out this explora-
tion, MM. Dollfus and Mont-Serrat — neither of them a geolo-

gist of reputation — spent a little over two years in that region, eight months of it in Central America. The results of their investigations have been laid before the public in the form of a ponderous quarto, in which, as in many other works of French savans which treat of the geology of parts of our continent, there is but little that is new, while, on the other hand, it contains many blunders. The Emperor has been unfortunate in the representatives of geological science whom he has sent to the American continent. M. Laur, who visited California some ten years ago, and made a report on its mines, showed a remarkable tact for misapprehending the plainest and most important facts, and drawing erroneous conclusions; as, for instance, when he announced that the yield of the Comstock Lode would never exceed three millions of dollars a year, whereas, in reality, it soon after reached twelve millions. About half of the volume of MM. Dollfus and Mont-Serrat is taken up with remarks on the volcanoes of Central America, and it is astonishing how little there is of original and valuable matter to be found in it. One is more annoyed still, on examining the beautifully engraved illustrations, to find that they bear evident marks of the sensational style ; the slopes of the cones are all enormously exaggerated, and no data are given by which these errors can be corrected. A few simple outlines plotted from actual measurements would have been worth more than the whole dozen and a half of costly steel plates which are given, the style of which takes us back to the dark ages of the illustrations to Humboldt's "New Spain." One should compare them with the drawings and sections illustrating M. Hartung's books on the Azores, Madeira, and Porto Santo, to see the difference between fancy and real work.

Baron Richthofen's quarto pamphlet of a little less than a hundred pages, with no illustrations, is entirely different from most of the works already cited, since it addresses itself exclusively to the professional geologist. It is the result of long observation and of much study bestowed on the volcanic rocks by an able and experienced observer in different parts of the world. In it many of the most difficult points in the theory of volcanoes are discussed in such a manner as to make its study imperative on all who desire to form an original opinion in regard to the subjects with which it deals. We shall refer to it further on, or at a future time, when the theory of volcanoes and earthquakes is under discussion.

In a previous article we endeavored to give a systematic view of the present condition of our knowledge of earthquake phenomena, so far as their external manifestations are concerned. We discussed the data of the earthquake catalogues with reference to the geographical distribution of seismic areas, to the relations of time of earthquake shocks, and to their connection with movements and conditions of the atmosphere. We had occasion to refer more than once to the relations between volcanoes and earthquakes both in time and space, and thus prepared the way for a discussion of the causes of these truly wonderful and most closely connected phenomena.

Before entering on this discussion, however, we must become more fully acquainted with the facts concerning volcanoes, and it is with these that this article will be occupied, leaving for a third and final one of the series an attempt to show how far science is able, at the present day, to throw light on those workings of unseen forces which are manifested in the earthquake shock, the volcanic eruption, the rising and falling of the land, and the formation of mountain chains, — for all these are effects of one and the same cause, or, at least, of one set of causes so intimately allied with each other that the discussion of any one of them must necessarily include that of all the others.

In pursuance of this plan, then, we purpose, in this article, to give an outline of what is known in regard to volcanoes, having reference chiefly to their external manifestations, such as form, geographical distribution, and their different phases of repose and action. This will prepare the way for us to get some idea of the nature of the forces at work below ; for a volcano is a sort of happy accident, which lets us into some of nature's secrets, — a peep-hole through which we may get a glimpse of the interior of the earth. It is evident that, if a great smelting establishment were buried so that no part of it should be visible except the top of the tall chimney, from which gases were issuing, and some piles of slags accumulated on the outside, and we had to report on the nature of the processes going on below from these imperfect data, the investigation would require no little scientific knowledge and ingenuity, and probably some time would elapse before a guess could be hazarded as to the character of the work of which these gaseous exhalations and slags were the only tangible result. So it is with volcanoes : we collect and analyze their products, whether solid, fluid, or gaseous ; we note the times and places of these manifestations of the internal forces and their correlations with other natural phenomena ; we avail ourselves of every conceivable source of information touching the subject, and reason to the best of our ability on the whole mass of evidence thus obtained. And yet the result, it must be confessed, is far from satisfactory. There are many obscure points in the theory of volcanoes and • earthquakes ; and if the general cause of the phenomena of volcanism is in the opinion of most geologists correctly determined, yet in regard to the precise mode of operation of the internal forces there is great discrepancy of opinion, even among those who have devoted most time to this branch of inquiry.

A volcano is a mountain, hill, or area of the earth's surface, connected with some more or less deeply seated portion of the interior by a canal or passage, through which solid or gaseous materials are brought to the surface. It is almost invariably the case that the substances thus ejected are intensely hot, the rocky material often pouring forth in a condition of igneous fluidity, and the term "lava" is applied to anything which has flowed in this way and which in cooling consolidates into rock. Elevations which would, according to the definition just given, be included under the head of volcanoes, but which emit only water with paroxysmal violence, are usually called "geysers." These are rare and on a small scale as compared with proper volcanoes. Orifices from which mud is thrown out, called "mud-volcanoes," are not uncommon, but are usually of small dimensions, and the temperature of the substances they eject is in many instances raised but little above their ordinary temperature.

Volcanoes are called "active" if they have within a comparatively recent period given indications of eruptive action. The term "dormant" may be used to designate that peculiar

condition when the internal forces have remained quiet for a
great length of time, so that only faint traces of activity are still
visible ; and if all chemical action has ceased, and there is no
record in history of any outbreak, the volcano or volcanic region is considered and called "extinct." Yet it is not an easy
thing to draw the line between dormant and extinct volcanoes.
Thus Epomeo, on the island of Ischia, remained entirely in-
active for seventeen hundred years. So Vesuvius was never
known in history as an active volcano until A. D. 79. A great
saucer-like depression, overgrown with wild grapes, in which
Spartacus once camped with ten thousand men, marked the
position of its crater, and Herculaneum and Pompeii were two
populous towns at its base. By the well-known eruption of
that year, these two towns were overwhelmed, — greatly to
the inconvenience of their inhabitants, no doubt, but immensely
to our advantage, — the whole adjacent region devastated, and
the mountain built up into an entirely different shape from that
which it had had before. From this time on, the eruptions continued, without any long periods of repose between them, until
the fourteenth century, after which there was quiet for nearly
three hundred years. During this period of repose the crater
became filled anew with a forest vegetation, and only a couple
of hot-springs gave evidence of the forces slumbering beneath.
All of a sudden, again, in 1631, a furious eruption took place,
and seven streams of lava flowed down the slopes of the mountain at one time. Since that, Vesuvius has almost always been
uneasy, there being rarely an interval of rest of more than ten
years, and, of late, the eruptions have been very violent and
frequent. The Gunung Gelungung, one of the great volcanoes
of Java, was, and had been from time immemorial, perfectly
quiescent, and the site of the present crater was a broad valley,
the inhabitants of which had never dreamed of anything but
the most peaceful security. But suddenly, in the middle of a
fine day in October, 1822, they received notice to quit, in the
form of a violent explosion beneath their feet, which proved to
be the commencement of one of the most fearfully destructive
volcanic eruptions on record.

There are but few volcanoes which are permanently active,
and those which are thus in constant eruption are usually far

from violent. Paroxysmal, powerful action occurs only occasionally, sometimes recurring, after short intervals, then slackening and perhaps ceasing altogether, or, after a long period of repose, say hundreds or perhaps thousands of years, beginning again.

We have in the moon the best possible specimen of thoroughly played out volcanism. The most careful watching of the surface with powerful telescopes seems, thus far, to have failed to reveal any evidence of changes taking place there. And since there is neither water nor air to produce erosion or disintegration of the volcanic surface, it seems pretty clear that it will remain as it now is for an indefinite length of time.

In dividing terrestrial volcanoes into extinct, dormant, and active, it must be understood, then, that these terms are used to express our general opinion with regard to their condition, based on a variety of circumstances, and not as indicating any positively established criterion by which the different classes can be distinguished from each other. We speak of the volcanic region of Central France as "extinct," because we know that a long time has elapsed since any indications of activity have occurred there; this has been ascertained by studying the amount of erosion which has taken place in the lava currents and in other ways. Yet the pouring out of a portion, at least, of the vast mass of volcanic material there visible took place, in all probability, after the appearance of man on the earth, although at an epoch immensely remote as compared with historical time. Neither can any conclusive reason be given why volcanic activity should not again manifest itself in this region.

A volcano may be considered as only dormant, and not extinct, when in the so-called "solfataric condition." This name is derived from the Solfatara, near Naples, where there has been no eruption since 1198, but where vapors and gases are constantly issuing from the region of the old crater. These vapors consist mainly of steam, mixed to some extent with sulphuretted hydrogen, and also with sulphurous acid, chlorohydric acid, carbonic acid, and nitrogen gases. The abundance of the sulphuretted hydrogen is usually testified to by the deposits of sulphur, so often met with in the craters of old

volcanoes, and undoubtedly formed by the decomposition of
this gas; besides, the nose has no difficulty, if no satisfac-
tion, in detecting its presence.   Steam and sulphuretted
hydrogen usually predominate largely among the products of
solfataric action.   The other gases mentioned generally, but
not always, occur in smaller quantity.   Boracic acid, petroleum,
specular iron, chlorides of the alkalies, realgar, and orpiment
are also occasionally observed among the gaseous emanations in
old volcanic regions.   Some observers testify to the existence
of inflammable gases in sufficient quantities to produce flames,
these gases being hydrogen and sulphuretted hydrogen; but
there are other observers, equally distinguished, who have had
frequent opportunities to examine volcanoes, both in action and
at rest, and who have never seen any indication of flame.   What
is generally called fire, in eruptions, is, of course, simply the
light or the reflection of the lava, which is intensely heated
but not actually undergoing combustion.

During the solfataric condition of a volcano, its crater becomes blocked up with congealed lava, perhaps overgrown with
forests and dense vegetation, and the signs of activity die out,
until, as the last relic of former life, only a thermal spring
may be found here and there, — an indication of the mighty
forces slumbering beneath.   Such is the present condition of
nearly all the great volcanic cones on our own coast, from
Arizona to Oregon.

Midway between the conditions of solfataric repose and of
paroxysmal violence is another stage of activity, in which some
volcanoes remain during long periods, while a few appear
never to pass out of it into more violent action; others, however, remain in this condition of partial repose during the intervals between violent outbursts.   At such times the crater
and the channel connecting it with the interior remain open, and
the lava can be seen in them maintaining a mobile condition,
while occasional explosions of the surface of the melted mass
take place, fragments of slag and cinders being thrown up and
mostly falling back into the abyss from which they were
hurled.   This was the condition of Vesuvius when visited by
the writer in November, 1843.   At that time there had been
no eruption of lava overflowing the lip of the crater since

1839, when the cavity was cleaned out, and left as a funnel three hundred feet deep, accessible to the bottom. From this time a smaller cone began to grow inside the large one, and in 1843 it was about fifty feet high, and could be reached by clambering down the walls of the old crater, the whole bottom of which, around the foot of the new cone, was covered with lava, which was red-hot a few inches beneath the surface, but could in most places be safely walked on. From the vent a shower of cinders was thrown up every fifteen or twenty minutes; and although it was possible to climb to the summit of the cone on the windward side, with occasional calls for skill in dodging the projectiles, the orifice was too much occupied with ascending vapors to permit of anything below being clearly seen. This interior cone kept on growing by additions made to it from the falling materials, and finally, in 1847, the crater became filled, and the lava overflowed, running down on three sides at once. From that time forward Vesuvius became very uneasy, and finally a great eruption took place in 1850. This lasted about twenty days, and when it was over the summit of the mountain was left much changed in form, the old walls having been broken down, the central cone reduced in size, and a new crater formed, about two miles in circumference, and a hundred and fifty feet deep. The volcano then remained quiet from 1850 to 1855, when it became very active; again a grand eruption occurred in 1858, and slight ones in 1860 and 1861. Since the last-named year Vesuvius has rarely been at rest. During the winter of 1867 – 68 there was a great outburst of volcanic force, which lasted several months.

In the condition of half-repose just noticed as not uncommon between intervals of paroxysmal activity, observers are able to look down into the throat or channel of Etna, as well as of Stromboli, during the periods of repose between the eruptions, which take place with great regularity every ten or fifteen minutes. At such times the lava is seen to move up and down in the chimney; as it rises, its surface swells up into a great blister, which finally gives way to the tension exerted and explodes with a loud noise, the fragments being scattered and thrown up with great force; the column of melted matter then sinks back into temporary repose, and rises again after an interval of a few minutes. The same phenomena were observed on Sangay, one of the Quito group, a permanently half-active volcano, like Stromboli.

The most gigantic exhibition of this condition of the volcanic forces is to be seen in Kilauea during its quiet periods, when the crater, which is three miles in its greatest diameter, has in it large pools of boiling and extremely fluid lava, which is continually thrown up in jets of from thirty to forty feet in height, that fall back into the pool before they have time to cool. These lakes of liquid fire vary in size according as the volcano is more or less active, and sometimes cover the whole area of the crater, the wind raising the surface in waves of molten rock, which dash against the encircling walls with an indescribably grand effect. The greater the liquidity of the lava, the less the force with which it is thrown up, for the jets of imprisoned vapors do not have time, in a very fluid material, to accumulate sufficient pressure to act with extreme explosive violence.

The phenomena which, we have seen, thus characterize the semi-active condition of volcanic activity are, in most respects, similar to those of the fully active state, differing rather in the degree of violence with which they are manifested than in kind. It seems, indeed, that the longer and more complete the repose of the volcano has been, the more violent its action when it once breaks out again. This is natural, for the resistance to an outburst must, as an ordinary thing, go on increasing the longer the vent remains stopped, and when this resistance is finally overcome the magnitude of the eruption will be proportionate to the force required to clear the way. The first recorded eruption of Vesuvius was the most violent of any which are known to have taken place; next to this in its destructive effects was that of 1631, occurring, as it did, after several hundred years of entire repose.

In regard to the precursors of a violent eruption, or the symptoms by which the approach of one may be detected, there is much uncertainty. It may be said, however, that a great outbreak is to be expected when the internal forces begin to show signs of uneasiness and the usual phenomena of half-repose to be intensified in their action. It seems a well-authenticated fact, that previous to an eruption of Vesuvius the wells and springs adjacent to the mountain begin to dry up. When volcanic cones are covered with snow it is not uncommon for the eruptions to be preceded by devastating floods, caused by its melting, the natural result of the gradual warming up of the mountain mass.

The following are the ordinary phenomena of violent eruptions : an appearance of fire ; lightning ; subterraneous noises, or thunder ; ejection of ashes, cinders, or blocks of lava ; the pouring out of melted lava ; and, in connection with earthquake shocks, fissures in the earth and permanent changes in the level of the adjacent country.

Great volcanic paroxysms are often preceded by more or less violent earthquake shocks, which are both frequent and prolonged, but usually limited to the mass of the volcano itself or its immediate vicinity. Tremendous underground detonations are heard, sounding like the firing of heavy cannon or repeated volleys of musketry. These sounds are heard at all points at the same instant of time, showing that they are propagated through the crust of the earth and also that they come from a great distance beneath the surface. These explosive sounds have been heard simultaneously over areas of many thousand square miles. Thus the noise of the outbreak of the eruption of Temboro, on the island of Sumbawa, was heard all over Java, and everywhere supposed to come from some point in the immediate vicinity. It was distinctly audible at points two thousand miles apart. As the shocks and sounds continue, people become more and more alarmed and excited, and imagine that they see every kind of portent in the sky or in the conduct of animals. It is generally thought that an oppressive stillness pervades the atmosphere just before the moment of the great outbreak, and that dogs, swine, and geese exhibit peculiar indications of fear. · How much reliance can be placed on the statements of the sensitiveness of animals to impending catastrophes, it is not easy to say ; but it is evident that the circumstances of a great eruption are eminently favorable to a highly imaginative condition of the mental faculties.

The earthquake shocks preceding volcanic outbreaks take

place while the internal conflict is going on between the imprisoned lava, seeking to find a vent, and the resistance offered by the weight and tenacity of the superincumbent crust. When the internal pressure which seeks relief in bringing up to the surface the material on which it is acting at last has its own way, the explosion is tremendous, the mass of the volcano being shaken to its very foundations. As soon as the channel of communication with the interior is opened, which channel usually communicates with the bottom of the old crater, although not unfrequently opened through some new side fissure, the pent-up vapors and gases begin to escape with tremendous force, carrying up in the air, torn into fragments, rocky masses, which then fall and are thrown out again repeatedly, and thus, by friction against each other or by actual explosion, through sudden changes of temperature, are rapidly reduced to powder and carried off with the gases or vapors which rise from the chimney of the crater.

The ejection of vapor and ashes, as the comminuted fragments of lava are called, is thus described by Scrope, who was an eye-witness of one of the grandest eruptions of Vesuvius, — that of 1822. He says: " The rise of the vapor produces the appearance of a column several thousand feet high, based on the edges of the crater, and appearing from a distance to consist of a mass of innumerable globular clouds of extreme whiteness, resembling vast balls of cotton rolling one over the other as they ascend, impelled by the pressure of fresh supplies incessantly urged upwards by the continued explosions. At a certain height this column dilates horizontally, and — unless driven in any particular direction by aerial currents — spreads on all sides into a dark and turbid circular cloud. In very favorable atmospheric circumstances, the cloud with the supporting column has the figure of an immense umbrella, or of the Italian pine, to which Pliny the younger compared that of the eruption of Vesuvius in A. D. 79, and which was accurately reproduced in October, 1822. Strongly contrasting with this pillar of white vapor-puffs is seen a continued jet of black cinders, stones, and ashes, the larger and heavier fragments falling back visibly after describing a parabolic curve. This jet of solid fragmentary matter often reaches a height of several thousand feet, while the vapor pillár rises still higher. Forked lightnings of great vividness and beauty are continually darted from different parts of the cloud, but principally its borders. The continual increase of the overhanging cloud soon hides the light of day from the districts situated below it, and the gradual precipitation of the sand and ashes it contains contributes to envelop the atmosphere in gloom, and adds to the consternation of the inhabitants of the vicinity."

If the volcano is one which emits lava, this rises gradually in the crater and finally overflows it at the lowest point, unless it succeeds in forcing its way through some side fissure. The molten mass finds its way down the declivity with a rapidity proportioned to its fluidity, overwhelming and destroying everything which it encounters. Clouds of vapor rise from the flowing mass, visible during the day, the exterior soon becoming covered with a dark crust of scoriæ, occasional fissures in which reveal, especially at night, the presence of the intensely ignited material beneath. The flow of lava from the volcanic vent indicates that the crisis of the disturbance is passed, and that there will thenceforth be a gradual slackening in the violence of the eruptive action.

Not a few volcanoes, however, never send out lava, but only ashes and cinders; these are usually the very large ones, as, for instance, the great cones of South America. It is also true that large volcanoes are less frequently than smaller ones the seat of great disturbances. The frequency of the eruption seems to be, in a measure, in inverse proportion to the height of the volcanic cones from which they proceed. Thus the lofty volcanoes of South America have rarely had more than one eruption each in a century; the Peak of Teneriffe had only three between 1430 and 1798. This is very natural, since the higher the cone the greater the resistance offered to an outbreak by the weight of the column. But the rule is not of universal application. Closely connected with the last-mentioned fact is another, previously suggested, namely, that the most fearful eruptions may be expected to occur after long intervals of repose. Both circumstances indicate very clearly the accumulation of force necessary to overcome increased resistance.

At night the column of vapor and ejected solid material appears red, not because it is actually a column of flame, but partly because it is illuminated by the reflection from the red-hot lava below, and also because the fragments carried up in it are themselves intensely heated. The fact that the column remains perpendicular all the time is a proof that it is not a flame, for, if that were the case, it would be swayed by the wind ; but one of the most characteristic features of an eruption is, that the pillar of fire seems to stand immovable amid the " wreck of matter " around it.

The electrical phenomena of a great eruption are extremely interesting. The upward rush of heated vapor gives rise to furious disturbances in the condition of the atmosphere, as is also the case, on a small scale, when steam escapes from an ordinary boiler through the safety-valve. A constant play of lightning goes on around the ascending column, and the noise of the thunder is mingled with the crash of the projected fragments of rock. Tremendous bursts of rain, or even hail, often occur at the same time, and from the same cause, — namely, the electrical disturbance of the atmosphere, — and the effect of the torrents of water rushing down the sides of the volcano is often more devastating than that of the lava itself.

The mass of ashes, scoriæ, or cinders thrown out in some volcanic eruptions is prodigious. In that of Vesuvius, in 1794, four cones were formed on a fissure nearly half a mile long, each with its separate crater, throwing up showers of red-hot cinders in such rapid succession as to appear like one continuous sheet of fire in the air. These showers really consisted of semi-fluid lava, which expanded in the air like soft paste. This continued for several days, so that the whole space above the crater seemed to be filled with the fragments, which formed a column a mile in circumference and rose to an immense height, then spread out, and seemed to cover a much greater area than the base of the mountain itself. Generally, however, these ejections of cinders are intermittent in character, sometimes following each other in rapid puffs, at others occurring as a succession of explosions at longer intervals.

The size of the fragments thus ejected is variable ; often they are as fine as the finest dust, but sometimes the lava is thrown out in great masses. Thus Cotopaxi vomited forth, in 1533, blocks of rock ten feet or more in diameter. The so-called volcanic bombs shelled out by Vesuvius are usually from the size of the fist to that of the head. Generally they are irregularly rounded or pear-shaped; but in volcanoes in which the lava is very liquid it comes down in masses which flatten out into cakes when they strike the ground. The finer fragments which in prodigious quantity accompany the larger, and usually vary from the size of a pea to that of a walnut, are now almost everywhere known by the Italian name of *lapilli*, or *rapilli*. The finer, sand-like material is called *puzzolana*, and the finest of all *ceneri*, or ashes.

One of the most curious features of the eruptions of some volcanoes is the prodigious number of small but perfectly formed crystals which are thrown out among the materials shot up from below. Vesuvius, which is a perfect treasure-chamber of interesting minerals, — while most of the American volcanoes are miserably provided in this way, — has furnished at times showers of beautiful crystals of augite, leucite, mica, and black garnet, the first-named being the most abundant. They seem to have existed ready formed in the semi-fluid lava, or else to have crystallized out suddenly at the moment of its solidification; which of these suppositions is the correct one is not thoroughly settled, although the first seems by far the most probable.

Vast masses of volcanic breccia occur in regions of eruptive rock, as for instance in California, where beds hundreds of feet in thickness are found covering many square miles of area, entirely made up of angular fragments of lava, of all sizes, which have evidently been ejected in the form in which we now see them. The explosions with which volcanic eruptions begin after long periods of tranquillity, and which sometimes pulverize the whole summit of the mountain mass in which they occur, give rise to prodigious accumulations of these broken masses of rock. The great eruption of Ararat, in 1840, was of this kind, a terrific explosion having torn open the side of the mountain and thrown off an immense mass of fragments, which were projected for miles in every direction, completely burying

4

the town of Argur6. There was no eruption of lava ; but frightful earthquakes and torrents of rain followed, washing down the
detritus of the explosion in immense floods of mud, which were
quite as destructive as lava would have been.

According to Junghuhn, the Javanese volcanoes now emit no
lava, but only give rise to streams of brecciated material, which
have issued from the craters in that condition.   The same author also gives a most interesting account of the great eruption
of Pepandayan, which took place in 1772. At that time such a
mass of fragments and blocks of lava was ejected that the upper part of the Garut valley, for ten miles in length, was filled
with ashes and angular materials to the average depth of fifty
feet, while in places the great blocks were heaped up in conical
hills as much as a hundred feet in height.   The distances to
which such masses are thrown indicate the immensity of the
force by which they are hurled into the air. Cotopaxi, for
instance, in 1533, threw rocks from eight to ten feet in diameter to a distance of seven miles.   The maximum height to
which masses of lava have been thrown by Etna and Vesuvius,
in different eruptions, is given by various scientific observers
as from seven to ten thousand feet.

Towards the end of an eruption the ashes ejected grow finer
and whiter, bearing all the marks of having been longer subjected to the triturating process by which the lava is reduced
to powder.   This is the natural result of the slacking off of
the ejecting forces, the sinking down of the column in the
chimney, and the consequent longer time that the materials are
exposed to friction against each other.   Some observers have
thought, however, that the lava might in many cases be blown
into fine powder by the sudden expansion into steam of the
water it contained, at the moment the pressure was removed
by its issuing from the crater, and there are some appearances
which seem to render this view a probable one.

The finer the ashes thus ejected, the farther away from
the volcano they fall. Carried by the wind, they are sometimes spread over vast areas of country, and the exceeding
fineness of the material is testified to by the slowness with
which it descends, sometimes filling the air so completely
that the darkness of night reigns for days in succession.   It is

stated that, in the great eruption which devastated the island of St. Vincent in 1812, the fall of ashes on the island of Barbadoes, nearly a hundred miles distant, caused so profound an obscurity that a white handkerchief was invisible at five inches from the eye. The fall of ashes in the great eruption of Temboro, in Sumbawa, in 1815, produced so dense a cloud that it was dark as night over the islands of Java and Celebes. Ashes fell on the islands of Sumatra, Banda, and Amboyna. West of Sumatra a layer of lapilli, two feet in thickness, floated on the sea, so that ships had difficulty in forcing their way through. A careful comparison of all the data, by Zollinger, led him to the conclusion that the ashes fell over an area of nearly one million of square miles, and that fully fifty cubic miles of material was ejected in this one eruption. Junghuhn, also, calculated the volume of the ejected materials of the same eruption to be one hundred and eighty-five times the dimensions of Vesuvius. The area over which daylight was shut off by this fall of ashes was nine hundred by seven hundred miles in extent, — that is, equal to the whole space in our own territory between the Mississippi River and the Atlantic Ocean. Coseguina, in 1835, covered with its falling ashes an area of nearly one thousand two hundred miles in diameter.

The destructive effects of these showers of ashes are fearfully increased by the torrents of rain which frequently fall in connection with great eruptions.; these carry down the ejected materials in the form of great flows of mud, which descend the steep slopes with such velocity that they cannot be avoided, and of course completely overwhelm everything they reach. It was by such a *lava d'acqua*, or water-lava, as the Neapolitans call it, that Herculaneum and Pompeii were submerged and destroyed. For eight days and nights the torrents of mud poured down over those ill-fated towns, accumulating in places to the depth of over a hundred feet. It was the remarkable way in which these cities were overwhelmed that has preserved them so wonderfully for the inspection of people for almost two thousand years. There is no other possible manner in which they could have been thus hermetically sealed up, as it were, all the walls remaining standing, and everything in its place. Had a shower of ashes, for instance, fallen from above, all the buildings would have been crushed in; but the insidious mud-flow crept into everything, filling rooms, and even cellars, so gradually that nothing was disturbed or displaced. Herculaneum was afterwards covered with a layer of solid lava, and then built upon, so that the opening of that town has been much slower and more expensive; although, in proportion to the amount of space uncovered, more interesting and valuable works of art have been disinterred than in Pompeii.

When these showers of ashes fall into the ocean, they gradually sink to the bottom, where they must eventually become consolidated into rock, which may be raised again to the surface in the course of the changes which are continually going on in the relative positions of sea and land. Thus are formed very extensive masses of stratified rock, which are at the same time both eruptive and sedimentary, or Pluto-Neptunean, as they have sometimes been called, as belonging to the two domains of the mythological rulers of the realms of fire and water.

It is by the constant addition made to their exterior by the falling masses of lava, ashes, and lapilli, that the cones of volcanoes are built up, — not only the dominating one of each volcano, but the secondary or minor ones, which are sometimes very numerous. These smaller cones form on the fissures which open frequently in the main cone, and which connect with the seat of action in the chimney of the volcano, just as that is connected with a still larger eruptive mass deep in the interior of the earth. Etna has more than seven hundred of these smaller cones around its base, some of which attain respectable dimensions, one reaching seven hundred feet in height, and another four hundred or more. On Vesuvius a fissure opened, in 1794, about nine hundred feet below the summit; this was two thirds of a mile in length, and eight new craters, with cones of scoriæ, were formed upon it.

Besides ashes and scoriæ, we expect, in most volcanic eruptions, to see rock rendered fluid by heat issuing from the crater, and it is to this molten rock that the name of lava is properly applied. The volcanic bombs, lapilli, and ashes are of course not fluid when ejected, although some of the larger masses sometimes reach the ground in a semi-plastic condition, so as to flatten themselves out into a sort of cake, as before mentioned. Different volcanoes and volcanic regions differ greatly in respect to the fluidity of their ejections. Those of Java, for instance, do not now throw out any molten lava, but only breccia, cinders, and ashes. The Hawaiian volcanoes, on the other hand, seem never to have ejected anything but lava of a high degree of fluidity. Vesuvius and Etna furnish both fluid and solid materials in abundance.

As a general rule, the very lofty cones do not emit currents of molten lava. Thus the great South American volcanoes throw out, almost exclusively, cinders and ashes. It may be stated that, in by far the larger number of instances, the great masses of lava which exist have come from low volcanoes, or still oftener from great fissures without any cones at all, in the form of " massive eruptions," as they are called, in which form, probably, by far the larger portions of the older volcanic rocks have come to the surface.

It is easy to see why, in lofty volcanoes, the lava should seek and find in many cases an issue at some point far below the summit. The higher the column, the greater the hydrostatic pressure, and when the resistance offered by this exceeds that which the sides of the mountain can oppose to it, the latter must give way, and the lava find a vent at the lowest available point. The constant battering of the internal walls of the chimney, kept up by the explosive forces within, gradually destroys the cohesive power of the material, breaks it up into fragments, or threads it in every direction with cracks, so that it finally yields to the repeated shocks, just as a piece of artillery fired with very heavy charges becomes at last too weak to resist any longer, and bursts into pieces. It is in this way that the fissures originate and become filled with molten lava, which solidifies in them, forming the dikes which are so common in volcanic masses, and which are so beautifully displayed in Etna, where its internal structure is revealed by the great cut into its heart called the Val del Bove.

The flow of lava, in volcanic eruptions, takes place in very different ways, according to its consistency and the position of the point from which it issues. In general the crater fills up gradually, until the fiery liquid rises high enough to pour over the edge at the lowest point, when it runs down the slope with a degree of rapidity proportioned to its fluidity. The Vesuvian lava is usually very thick and ropy. One of the greatest currents of that volcano, — that of 1794, — which was over a thousand feet broad and from twenty to thirty deep, ran two and a half miles in six hours, or at the rate of 2,160 feet in an hour. The lava of Mauna Loa, on the other hand, is so liquid, that when it issues from the crater it pours down the steep slope of the mountain, sometimes with amazing velocity. Thus Mr. Coan says of the eruption of 1855: "In one place only we saw the river [of lava] uncovered for thirty rods, and rushing down a declivity of from ten to twenty-five degrees. The scene was awful, and the momentum incredible. The fusion was perfect, and the velocity forty miles an hour." This lava, in making its way down the mountain-side, leaps over precipices in literal cascades of fire, presenting a most sublime spectacle. It occasionally forces its way out from a side fissure, — under immense pressure of course, — when it plays as a fountain, and the jets of liquid fire are reported by trustworthy authorities as rising sometimes to the height of six hundred or eight hundred feet.

Lava streams, however fluid the material may be, soon become covered, as they run down the sides of the volcano, with a consolidated crust. This hardened surface gradually thickens, and the bottom and sides also become more or less congealed, so that the flow continues through a sort of tunnel, as if it were being poured out of a sack made of its own substance. The surface gets broken up into great angular masses, which, by the motion beneath, are thrown into disorder and piled up on each other, as cakes of ice are on the sudden breaking up of one of our great rivers, — the St. Lawrence for instance. In the great eruption of Mauna Loa, already mentioned, the lava made its way seventy miles reckoned by the course of its flow, and forty in a direct line, to Hilo; and after its surface had become quite hard all the way, and there was no evidence of activity visible except the columns of vapor ascending from its head and foot, Mr. Coan believed that the interior was still moving downwards. This stream of lava was three miles wide on an average, and in some places three hundred feet deep. The masses of broken crust were piled up on it to the height of a hundred feet at various points.

The lava of Vesuvius seems more variable in its consistency than that of almost any other volcano. In the eruption of 1805, the velocity with which it issued from the crater was almost equal to that of the Mauna Loa current; on the other hand, the stream of 1822, when it reached Resina, moved at the rate of only five or six feet an hour. That of 1819 was in motion, at the rate of three feet an hour, nine months after its issue. The rate of motion, measured by Dolomien, of one stream, was a mile a year.

The low conducting power of lava is the reason why the interior of the mass can remain fluid so long and run beneath a crust of its own substance. The exterior hardens, can be walked over, or perhaps even cultivated, while the interior is still red-hot. This internal heat lasts for a long time. The lava of Jorullo was hot enough to light a cigar twenty-one years after its issue; and sixty-six years later it was still perceptibly heated, sufficiently so to give rise to *fumaroles.* One of the lava flows of Etna — that of 1787 — spread over a mass of snow, which, in 1830, still remained under it unmelted, while the overlying mass of rock was quite hot. The snow was preserved from melting by a cover of ashes, through which the heat was conducted with extreme slowness.

The manner in which volcanoes are built up by successive ejections of ashes, scoriæ, and lava, and the question whether the vast size of some cones is due in part to any other cause than this simple one of the piling up of erupted materials around a central orifice, now remain to be discussed.

The simplest possible form of a volcanic accumulation is that of the ordinary cinder cone, built up by a single eruption. Such cones are among the most common, as well as the most characteristic, features of almost every volcanic district. The coarse fragments thrown out heap themselves around the orifice as they fall, in the form of a circular bank, which, as the eruptive action continues, increases in size until it becomes a hill, having the form of a truncated cone, with a funnel-shaped hollow at the summit. A section of one will show that they are rudely stratified, and that the inclination of the strata decreases with the distance from the centre. These cones are of all sizes, from that of a hay-cock to that of a mountain. The "Puys,"

as they are called, of Central France, — Auvergne, Velay, and
the Vivarais, — are hills of scoriæ thrown up in this way.
Near Clermont-Ferrand there are above sixty cones strung
together on a line more than sixty miles in length, and the fissure on which these were built up is continued in Velay and
the Vivarais, with two hundred or more such cones arranged
in a belt twenty miles long. The shape of these accumulations
of ejected materials varies with the conditions under which
they are formed. When the wind blows steadily from one
quarter, the materials will be more heaped up on one side,
and this effect is very marked in the region of the trade-winds.
A great many causes may be effective in modifying the cones
thus formed. One is the issuing from them of a current of
lava, by which the mass is broken down on one side ; such
breached cones are among the most common features of many
volcanic regions.

An ordinary cone resulting from a single eruption consists,
then, of a pile of scoriæ, lapilli, and other loose materials, with
a single current of lava, which may have flowed from the summit, the side, or the base of the elevation, and which will be
found spreading itself out over the adjacent region in a sheet or
stream, proportioned in size to the extent of the eruption and the
nature of the surface over which it has found room to extend itself. The result of repeated eruptions occurring from the same
vent will be the gradual building up of a mass, which grows in
size constantly, but has the same kind of structure from top to
bottom. Beds of solid lava alternate in it with others of fragmentary materials, and the whole system dips in all directions
from the centre. It is not to be supposed, however, that any
one of the beds of lava entirely surrounds the cone ; on the
contrary, if a horizontal section were made through such an
accumulation, it would be seen that each outflow of molten
rock has only added to the mass a portion of a concentric belt,
so that the cone is built up by gradual additions of ejected materials, first on one side and then on another. Besides, there
would be found, in many cases, a net-work of dikes of lava
ramifying through the lower interior portion of the cone, and
produced in a way which has already been indicated.

Almost all the older authors and many modern ones suppose that all volcanic cones have been built up in this simple manner. The theory originated by Humboldt and elaborated by Buch, and called the "crater-of-elevation theory," has found many warm supporters even among those who have worked long in volcanic regions, while it has been persistently opposed by most of the English geologists, especially by Lyell and Scrope, as well as by Dana in this country. According to this theory, most great volcanoes consist of two portions, very distinct from each other in their mode of formation. The lower part, or base of the mountain as it might be called, consists of strata inclined at a less angle than the upper, and has not been formed by the accumulation of ejected materials, but consists, rather, of stratified masses, which may have been sedimentary beds deposited horizontally, or volcanic materials erupted from fissures under the ocean. In either case these beds are supposed, by the upholders of Buch's theory, to have been brought into their present inclined position by a "bubble-shaped elevation of the ground," caused by pressure of the volcanic forces confined beneath. On this inflated mass, through the centre of which the lava afterwards found its way, the cone of eruption is supposed to have been formed in the ordinary manner. In many cases, however, the process was a more complicated one. After the formation of the flattened dome-shaped mass, the volcanic energy, exerting itself at the base of the chimney by which the dome was penetrated, would fracture it in all directions, force lava into these fissures, swell out the mass, and gradually open a great crater at the summit, around the edge of which the strata would stand at a much greater angle than they originally had, it being maintained by Buch and the upholders of the elevation theory that lava could not consolidate in thick beds on steep slopes, — an assertion which has been abundantly disproved by observations in different parts of the world.

It is in this condition of dome-shaped elevation, caused by pressure from beneath, that Vesuvius is supposed to have been at the time Spartacus camped in its crater, just before the great eruption of 79. At the time the explosion took place, and the hidden forces obtained an outlet, one side of the crater of elevation was blown off, and an ordinary ash and cinder cone began to form in the cavity. The same mode of formation is claimed for Etna by Élie de Beaumont, one of the most zealous supporters of Buch's theory, who maintained that the lower portion of this great volcano was quite distinct in its formation from the upper ; that the one was formed beneath the sea by the elevation of horizontally deposited strata, while the other, or the cone proper, — which is eleven hundred feet high and has as steep an angle as thirty-two degrees, — was built up by subaerial accretions exclusively.

Buch applied his theory to the Peak of Teneriffe, of which he made a most detailed examination, and endeavored to explain by it the formation of the great semicircular wall which encloses the peak itself and the cone of Chahorra. This encircling precipice is, in places, full two thousand feet high and no less than eight miles in its longest diameter. Buch also visited and described with minuteness the beautiful island of Palma, a little west of Teneriffe, which is another of these great truncated cones, with a huge and deep cavity in the centre, called by the natives a *caldera* (kettle), from three to four miles in diameter, and walled in by a precipice varying from fifteen hundred to twenty-five hundred feet in vertical height. This boundary wall is so steep and unbroken that there is only one place where a descent is possible even on foot.

This kind of structure — namely, an encircling ring, of enormous dimensions compared with those of ordinary craters, with a cone in the centre — is quite common, and is especially well seen on some volcanic islands, where the internal structure is revealed by breaches made by the sea in the exterior wall. The interesting island of Santorin, in the Grecian Archipelago, is a good instance of this kind of arrangement, the volcanic fires here having been active of late, and the region one which has furnished material for a considerable number of volumes, as already mentioned. The island of Nisyros has a similar structure, the nearly circular crater being three miles in diameter and surrounded by a rim which rises from two thousand to twenty-three hundred feet above the sea. The island of St. Helena is described by Mr. Darwin as a trachytic volcano, encircled by a broken ring of basalt, measuring eight miles in diameter one way and four the other; the internal cliff faces are nearly perpendicular, except that they have in some places flat projecting shelves or ledges cut around them in parallel curves. Barren Island, in the Bay of Bengal, and the Mauritius, are other excellent examples of the same interesting type of structure. The encircling crater ring of the last-named island measures no less than thirteen miles in diameter.

Lyell, in the tenth edition of his " Principles of Geology," published two years since, has gone pretty thoroughly into the question of the applicability of Buch's theory to both Vesuvius and Etna, giving the results of his own repeated and recent examinations of these classic volcanoes, and pointing out that many important facts had been misapprehended by those geologists who had endeavored to show that the crater-of-elevation theory was the only one applicable to explain their form and structure. Hoffmann, many years ago, after a careful study of Vesuvius, abandoned the theory of Buch, which he had previously maintained. Of eminent French geologists, Cordier and Constant Prévost were also opposed to the idea of the building up of volcanoes in any other way than by the piling of one layer of ejected materials upon another.

The principal difficulty which those who do not support the crater-of-elevation theory have to meet is the enormous size of some of these great encircling rings, which would seem at first too large to be the result of explosive forces, implying as they do an astonishingly violent action and areas of vast dimensions over which the volumes of vapor must have been driven upwards.

There are craters of gigantic size, however, in regard to which it seems clearly demonstrated that they were formed in the ordinary way, that is, by the aggregation of materials erupted from a central orifice. Thus Kilauea does not bear any marks of being a crater of elevation; neither does the grand Haleakala, on the island of Manui, which is estimated to be some thirty miles in circumference. Junghuhn, who has made such a careful examination of the volcanoes of Java, gives it as the result of his observations that the great cones of that island have all been formed by eruption, and not by elevation; and he gives most excellent reasons for drawing this inference,— such reasons, indeed, as could only be successfully opposed by proving him to have misstated the facts. Similar conclusions have been arrived at by the writer of this article, after examining several of the great cones on the Pacific coast of North America.

If we consider what prodigious masses of material are thrown out, as already mentioned, in such eruptions as that of Temboro or Coseguina, it will not be difficult to understand that a cavity of corresponding size must be left behind ; and, as a means of enlarging such a cavity to an almost indefinite extent, we may call in both subaerial and submarine erosion, although the former has probably been usually by far the most effective agent in this respect. That any such great blister-like uplift of the superficial crust as was imagined by Humboldt to account for the dome-shaped base of Jorullo ever occurred seems, on the whole, highly improbable. His idea of a hollow crust or roof blown up over a vast empty space beneath will hardly be adopted by any geologist at present. Everything indicates, on the contrary, that, instead of there being a vacuum or a space filled only with gaseous substances under or over the centre of the volcanic action, there is much more likely to be a crowding together in that region of fluid material, seeking to find a vent. That great areas of stratified deposits might, under such conditions, be elevated into dome-shaped masses, is certainly not impossible ; and yet it is questionable whether the fact of any such occurrence has ever been demonstrated.

It is indeed curious that the great name of Buch — a man once the very leader of geological science, and to whom Humboldt dedicated his *Kleinere Schriften* in these words : " Dem geistreichen Forscher der Natur, dem grössten Geognosten unseres Zeitalters, Leopold von Buch " — should for many years back have been most frequently quoted in order to bring forward fresh evidence against some one of his favorite theories, or to show how thoroughly he misapprehended some great geological phenomenon, like that of the distribution of the glacial boulders in Switzerland. Still the fact, however discouraging it may seem to those looking simply to permanence of personal reputation, is, in reality, an indication of progress in the science. Had Buch made a thorough examination of the geologically classic region of Southern Tyrol, he never would have given to the world a theory so entirely unsupported by facts as that by which he sought to explain the formation of the wonderfully picturesque cliffs of dolomite which have made that country so celebrated, and the origin of the rock of which they are composed. The day of generalizations of a magnitude entirely disproportionate to the slender base of facts on which they rest has passed away ; or, at least, the practice of bringing such theories forward with the positiveness, and upholding them with the obstinacy, of a Buch, is one which is no longer in vogue.

There are, indeed, many geological phenomena the theory of which is obscure and difficult, and for whose final elucidation the stock of accumulated observation is still insufficient. If, with the view of directing attention to deficiencies in this stock, rather than of parading his actual knowledge, the geologist groups these facts together, and endeavors to show in what direction they seem to point or what the ultimate solution of the problem will probably be, he will, if his work be done in the right spirit, not incur the charge of rashly generalizing or of endeavoring to force his opinions on others. Among the obscurest and yet most attractive topics of geological investigation it would be safe to include the theory of volcanoes and earthquakes, and especially the connection of their phenomena with those movements of the earth's crust which have resulted in the formation of continents and mountain-chains, and which, by altering the relative level of land and sea, have played the principal part in the long series of events that have been going on since our planet became the theatre of geological changes. This article, and one in the preceding number, may be considered as leading the reader to a point from which he will be able, with profit, and, it is to be hoped, not without pleasure, to survey the indicated field, and we shall endeavor at a future time to act as his guide in such a survey. Before closing, we must add a few pages to what has been said in a previous article, in regard to the geographical distribution of volcanoes, or their arrangement upon the earth's surface.

By far the most interesting fact in this connection is the

proximity to the ocean of almost all active volcanic vents. Probably nine tenths of them are distributed around the Pacific, forming what has been aptly called a "circle 'of fire" full twenty thousand miles in length. The islands on the west side of that ocean form almost a continuous chain, beginning with the Aleutians on the north, and extending to New Zealand on the extreme south. This is pre-eminently a region of active volcanism, for hardly a single one of the numerous islands in the various groups of which this belt is made up is entirely destitute of active vents, while on some of them they are crowded together by the hundred. In the groups of the Formosa, Philippine, Molucca, and Sunda Islands, there is perhaps the greatest concentration of volcanic energy which our planet exhibits. Nor is the east side of the Pacific less bountifully supplied with indications of igneous activity. Along the whole coast, from Patagonia to Alaska, the eruptive formations are displayed on the grandest possible scale, although the regions of present activity are sometimes widely separated from each other, and the volcanic belt, taken as a whole, presents evidences of a very considerable slackening of its energy since the close of the Tertiary period.

In the South American Andes the active volcanoes are chiefly limited to three great systems,—those of Chili, Bolivia, and Quito. Each of these has its grand cones, among which are the highest points in the world, with the exception of a few in the Himalaya. Aconcagua, the monarch of the Chilian group, lacking not much of twenty-three thousand feet in height, has been generally supposed to be a volcano, and was even reported by Darwin as having been in eruption in 1835. Some doubts have been thrown on this statement, however, by M. Pissis, a topographical engineer, who has been employed for years by the government of Chili in making a map of that country, and who maintains that Aconcagua consists of rocks of the Cretaceous series. It is curiously indicative of the feebleness of the spark of scientific inquiry which is kept alive even in the most enlightened of all the South American states, that so interesting a question should not have been definitely settled a long time ago. Still higher than Aconcagua is Sahama, chief of the Bolivian group, and only surpassed in elevation, on the American continent, by Illimani and Illampu. It is twenty-four thousand feet high, or one thousand feet higher than Chimborazo, which was long supposed to be the most elevated mountain mass of the New World, but which, although the loftiest of the magnificent group which surrounds the plain of Quito, is only 21,420 feet in height. Off the coast of Central and South America, at a considerable distance, however, are groups of volcanic islands, with long intervals between them, which may be compared with the similar, but far more closely crowded ones on the opposite side of the Pacific. Along the line of these groups, within the intervals between them, frequent volcanic submarine eruptions have been observed, which have given rise to islands; these, however, have since been mostly washed away. If we may judge of the future by what has occurred in the past, it would be safe to predict that, as volcanic action dies out on the present coast line, a new belt will be gradually added to the continent on the west side. We might, without being considered as indulging in a fanciful speculation, say that the process of adding such a belt on the Asiatic side was already far advanced, while on the American it is just beginning. The most remarkable instance of insular volcanism on the east side of the Pacific is the group of the Galapagos, five hundred miles off shore, in the latitude of Quito. This group consists of five principal islands and several smaller ones, all volcanic. Craters have been seen in eruption on two of these, and on several of the others the streams of lava have quite a fresh appearance. The number of craters on the group is very great, having been estimated by Darwin at as high a number as two thousand.

The volcanic phenomena of the west coast of North America are on a still grander scale than those of the southern half of the continent, as far as the extent of the area covered by igneous products is concerned. There are not, however, as many very lofty cones, and not, in general, as much present activity. The highest development of volcanism on that coast seems to have occurred just at the close of the Tertiary epoch, and at that time the activity of the internal forces must have been prodigious. In spite of the immense erosion which has

taken place since that time, the proofs of this activity are everywhere visible along the whole line of the coast from Central
America to Alaska. The regions of active volcanic excitement
on the Pacific coast of our continent are at present but two in
number, and these are placed at the two extremities of the
line, one in Central America and Southern Mexico, the other in
Alaska and the Aleutian Islands. The southern region is
divided into two groups, the Central American and the Mexican; the former begins with the volcano of Chiriqui and extends to that of Soconusco, on the Isthmus of Tehuantepec, —
a distance of full eleven hundred miles. This group is remarkable, not only on account of its parallelism with and close
proximity to the coast, but for the number and size of the
cones of which it is made up; of these there are more than
fifty, almost all on the summit or else on the western flank of
the Cordilleras. Perhaps, with the exception of Java, there
is no region in the world where the volcanic vents are so
crowded together. Of all the eruptions which have taken place
here during the historical period, that of Coseguina, in 1835,
already mentioned, was the most astonishing. The ashes
thrown out at that time produced darkness for two days over a
great extent of country, and covered an area as large as that
of New England to the depth of several feet, the noise being
heard in Jamaica and at Bogota.

Four hundred miles north of Soconusco, and exactly in a
line with the prolonged axis of the Central American volcanic
belt, rises the cone of Popocatapetl, generally considered the
loftiest point of North America, and certainly the highest which
has been accurately measured. Its only possible rival is its near
neighbor, Orizaba, which has been made by some late, but not
very trustworthy, measurements a little the higher of the two.
Popocatapetl has been repeatedly measured with closely coincident results, so that we probably know its height within twenty-
five feet; it is about 17,750 feet. Both these great cones belong
to the chain of lofty volcanic vents which traverses the continent, in the direction of east and west, nearly in the latitude of
the city of Mexico. Beyond this belt to the north, within the
limits of Mexico, there are no active volcanoes; nor are there
any on the peninsula of Lower California, as is uniformly

reported in all the books; there are but few volcanic cones even, although rocks of this character in the form of dikes and sheets of lava are abundant in some parts of the peninsula. The volcanic formations on the mainland opposite are extensive and wonderfully varied in character; but they all belong to a past epoch of activity.

Crossing the Mexican boundary, and entering our own territory, we find eruptive rocks abundant; and, on reaching the parallel of 35°, a little to the north of the centre of Arizona, another great volcanic belt may be traced across the Cordilleras, in a line transverse to their general trend. The most prominent cones of this belt are Mount Taylor, San Francisco Mountain, and Bill Williams's Peak, all magnificent mountains, probably between twelve and fourteen thousand feet high, but none of them has been ascended or accurately measured. They rise grandly from the plateau of horizontally stratified rocks, and are surrounded by vast lava fields bearing all the marks of having been erupted at no very remote period, although there are no indications of present activity.

Passing up through California and Nevada, we find all along both slopes of the Sierra Nevada, and on the parallel ranges east, entirely through to Salt Lake, abundant evidences of former volcanic action, on the grandest possible scale. On the east side of the mountains, this condition of activity seems to have ceased at the commencement of the present geological epoch, or at least to have diminished greatly in violence. The only indications of present volcanic activity along the Sierra Nevada, south of the north line of California, — aside from the numerous hot-springs, — are some comparatively faint remains of solfataric action on a few of the highest points. Thus Lassen's Peak, for instance, has several quite large areas where sulphurous gases escape from pools of hot water and boiling mud, while near the summit of Mount Shasta, amid the eternal snow, there is a hot-spring from which sulphurous vapors are constantly issuing. Between these two lofty volcanoes, one nearly 11,000 and the other 14,440 feet high, there are many others, some with wonderfully well-preserved craters, looking as if of very recent formation, yet entirely destitute of any traces

5

of present activity. On the eastern slope of the Sierra, near Mono Lake, are a number of lofty and beautifully regular cones with well-defined terminal craters, yet apparently quite extinct. All through the State of Nevada, indeed, the mountain ranges are extensively flanked by vast accumulations of lava, and when we cross the Humboldt River, and traverse the region north of the parallel of 41°, we find a continuous covering of volcanic materials extending over all the northern portion of Nevada and California, as well as Southern Idaho, Eastern Oregon, and Washington Territory. This region, which is covered almost exclusively with basaltic lava, is but little, if any, less than six hundred miles square, and occupies an area considerably larger than France and Great Britain combined. It is by erosion of rocks of this character that the many beautiful waterfalls of the Snake, Pelouse, and other rivers have been formed. Those of the Snake River are described by the few who have seen them as of surpassing grandeur. They must be among the very finest in the world, taking into account height, volume of water, and attractiveness of the surrounding scenery.

North of the California line the belt of nearly extinct volcanic activity is continued in the Cascade Range, — the prominent peaks and cones of that chain, which is in fact a continuation of the Sierra Nevada, being all of volcanic origin. The best known ones south of the Columbia River are — naming them from south to north — Mount Pitt, the Diamond Peaks, the Three Sisters, Mount Jefferson, and Mount Hood. The latter is a magnificent cone, very conspicuous over a great extent of country, and much looked up to and respected by the Oregonians, who were very wroth at having its boasted 17,000 or 18,000 feet cut down by the ruthless hand of science to 11,225. North of the Columbia are Mount Adams and Mount St. Helens, which are in nearly the same parallel; then, Mount Rainier, standing in solitary grandeur about seventy miles east-south-east of Olympia; and finally, Mount Baker, near the line of British Columbia. Of these great cones, Mount Rainier is the noblest: as seen from Puget's Sound, covered with snow nearly down to its base even late in the summer, it is truly a magnificent object. Its summit has never been reached, so far as we can ascertain, while all the other important cones of this region have been repeatedly ascended. That any of these volcanoes have emitted streams of lava since the country became known to the whites is not probable; but that ashes have been thrown out from two of them, Mount St. Helens and Mount Baker, seems to be well authenticated. The newspapers have frequent accounts of columns of vapor being seen to issue from Mount Hood, and of other indications of activity being displayed by the great cones which are such conspicuous objects to those passing up and down the Columbia. These stories, when not intentional fabrications, may perhaps be attributed to the fact that sometimes on clear days the moisture in the air blowing from the ocean is condensed around the cool, snow-covered summits of the cones, so as to have somewhat the appearance to a not very critical eye of clouds of vapor issuing from them. We obtained pretty satisfactory testimony that Mount Hood at least had shown no signs of activity during the past eight or ten years.

There are also most conflicting statements with regard to the condition of the volcanoes through British Columbia and Alaska. Thus Scrope, a careful and trustworthy authority, says of Mount St. Elias, that it has certainly been seen in eruption, while Grewingk, a well-known geologist who explored that region and carefully examined all the published authorities on the subject, declares that none of these volcanoes — St. Elias, Edgecombe, Fairweather, etc. — have been active during the historical period, or, at least, that there is no evidence of any such activity.

VOLCANISM AND MOUNTAIN-BUILDING.

In two articles published in previous numbers of this Review (Vols. CVIII. page 578, and CIX. page 231) we have discussed the phenomena of earthquakes and volcanoes, endeavoring to convey in popular language some idea of the nature, extent, and frequency of these remarkable manifestations of the internal forces of the earth. In the last of these two articles it was suggested that occasion would be taken to continue the consideration of the subject, and to endeavor to explain, or at least throw some light on, the nature and connection of the chief causes which have been concerned in carrying on that complicated series of geological dynamics which we include under the comprehensive term " volcanism," and of which the earthquake and volcano are two of the most striking manifestations. The subject is one which has always commanded the attention of geologists, and suggested, or even provoked, discussion among them. The difficulties which it presents, however, become apparent, when we learn, through examination of the printed records of these discussions, how little agreement there is among geological authors, and how widely they differ in regard to points which, as one would suppose, ought long since to have been settled.

We have repeatedly, in the course of the preceding articles, referred to the intimate relationship existing between the phenomena of earthquakes and volcanoes, — a relationship which can hardly fail to have become apparent to all who have given even a limited amount of thought to the subject. We have now, however, to go one step further in the same direction, and show how the consideration of the subject of volcanism leads most directly and naturally to that of the formation of mountain chains, or, in still more general language, to the study of

6

the forces which have drawn the outlines of the continents and
moulded the surface of the earth into its present relief.

The difference of elevation between the summit of the highest land and the bottom of the deepest ocean is but trifling, as
compared with the whole diameter of the globe, yet of what
immense importance in the economy of nature are our mountain chains, and how thoroughly are our most weighty interests
and avocations dependent on the form and elevation of the
continental masses! Mountains as geographical and geological facts are of the highest significance to the student of
natural phenomena, in whatever light he considers them. As
agents in determining the character of the climate, the courses
of rivers, the nature of the soil, the migrations of nations, the
distribution of languages, manners, and customs; in short, in
their relations to man and nature, from the point of view of
physical geography, they play the leading part. As permanent
records of past geological changes in the history of the earth,
mountains are all important to the student of geology. What
would this branch of science be, without mountains and the
study of mountains! The results of the dynamics of the globe
are registered in the mighty ridges which encircle it, and mark
the outlines of its continents and oceans. Indeed, we can
hardly conceive of the present order of things as existing at
all without these visible results of the manifold causes which
have worked together to make the earth habitable.

Hence, the study of the structure and mode of formation of
mountains is the study of the greatest problems with which the
science of geology presents us. Thoroughly to work out and
comprehend the structure of all the mountain chains of the
world would be little different from thoroughly working out
and comprehending its geology. There is hardly a problem
presented by the science which would not find its application
in some one of our mountain systems. Orography, then, or
the study of the structure of mountain chains, is the study of
geological phenomena on the largest scale and from the most
generalized point of view.

It cannot fail to have been impressed on the mind of the
reader of the preceding articles, that there is an intimate connection, in the character of the results at least, between the

forces by which volcanic and earthquake action is kept up and mountains originated. A volcano is, in fact, a mountain, and to the popular mind there is little difference between an isolated elevation and a group or line of them; between a mountain and a chain of mountains, in short. But not a few of the latter are almost or quite exclusively aggregations of volcanic material; and when we come to rocks which are eruptive, that is, which have been poured forth from the interior of the earth, although not technically volcanic, we find that these constitute a large portion of a great many mountain chains, and especially of the highest and grandest. And the more we look into the matter the more we shall be convinced that the formation of mountains and the development of continents are also closely correlated phenomena. Mountains are but the skeletons of the continents. Wherever a lofty chain of mountains has been raised above the sea-level, there is a central mass with a tendency to grow and spread itself laterally, under the influence of denuding agencies; and, unless counteracted by a general subsidence, there will be a steady increase of breadth of the region, at the expense of the height of the more elevated portion. If the material is carried down and deposited under the ocean, then, whenever there is a change in the relative level of sea and land, so as to bring the newly formed strata above the water, these will be found to present evidences of the conditions under which they were deposited, in the form of marine fossils, which will be more or less abundant, according as the physical conditions varied at the time the deposition took place. If two chains of mountains are so situated with respect to each other that continental growth may take place between them, the process will, of course, be the more rapid, and the newly made land will cover a proportionably greater area. Every continental mass, then, will be found on analysis to consist of one or more chains or groups of mountains, and large areas of lower land which has been derived from the long-continued erosion of the more elevated regions. An examination of good topographical maps of the different continents will show this relation most clearly; especially if aided by sections across the land-masses transverse to the direction of the leading chains of mountains which traverse them.

It is clear, then, that when looking at the subject from the broadest point of view, and endeavoring to make out what agents of geological change have been most widespread and general in their action, we cannot separate the phenomena of volcanoes and earthquakes from those of mountain-building and continental growth. One key must give access to all the mysteries of geological dynamics. The nature of this key was first rather vaguely shadowed forth by Leibnitz in his "Protogæa"; but the key itself was not really forged until long after, when Humboldt began to group the physical phenomena of the universe into one harmonious picture, or cosmos. Leibnitz recognized the fact that the earth had cooled from a condition of igneous fusion, and that in this cooling inequalities of the surface would be likely to have arisen. But it was reserved for Humboldt to announce a cause of volcanic action which would be always operative, both through the past geological ages and in all future time. As first enunciated by him, half a century ago, it was intended to be applied solely to volcanic phenomena, and was thus expressed: "All volcanic phenomena are probably the result of a communication, either permanent or transient, between the interior and exterior of the globe." Ten years later, the idea of one general cause for all the varied forms of volcanism has clearly worked itself out in Humboldt's mind, and was thus formulated in the "Kosmos": "In one comprehensive view of nature, these all (namely, the phenomena of volcanism) fuse together into the single idea of the reaction of the interior of our planet against its crust and surface."

While now most geologists admit the validity of this explanation, so far as it goes, the discrepancies of opinion which have arisen in showing how the reaction in question is brought about are very considerable. As long as the theory was only vaguely shadowed forth, and no attempt was made to go into details, but little objection could be offered to it. But when, as facts began to accumulate and more precision of statement and clearness of development were required, in harmony with the progress of modern thought, the difficulties of the case became more and more apparent and the divergencies of opinion greater. Humboldt may be said to have furnished a blank key, which looked, at first, as if it would fit the lock; but every examination has revealed some new ward to which it must be adapted; and different observers have shown themselves very much in doubt as to how it was to be filed to fit the complications which they had discovered, and which combine to make the opening of the lock anything but the simple task which it seemed at first to be.

As time passed on, and the various borings and mining operations all over the world gradually gave absolute certainty to the at first rather hesitatingly received fact of a universal increase of temperature in descending beneath the earth's surface, the views of Humboldt began to be generally received and acquired something like this form : The earth is gradually cooling from a condition of intense heat and igneous fusion. During this cooling an exterior crust or shell has been formed. This crust has, from time to time, been endeavoring to adapt itself to the still shrinking nucleus, and, while so doing, has from time to time yielded to the accumulating tension. The vibration resulting from this sudden yielding has been the principal cause of earthquake shocks, and through the fissures thus formed the molten matter of the interior of the earth has, at various intervals, found its way to the surface in the form of volcanic eruptions and accompanied by all the phenomena of volcanic action. The crust of the earth, in endeavoring to adapt itself to the nucleus, has been in places more or less uplifted or depressed, folded or plicated, thus giving rise to those irregularities of the surface which we call mountains, and which, also, often owe their existence to a direct pouring out of the eruptive material through an elongated fissure, this material then forming the axis of the mountain mass or range.

Those who are familiar with the various geological text-books in use will recognize that this is the simplest way of expressing the generally adopted theory ; but, as will presently be seen, there is the widest variety of opinions and hypotheses when anything like an approach to a detailed statement of the *modus operandi* of the internal forces is attempted.

In the first place, there is considerable diversity of opinion among geologists as to the manner in which the earth has consolidated while cooling. We know that the specific gravity of

our planet, as a whole, is about double that of its external
crust; and to account for this superior density of the interior,
we have to endeavor to combine three conditions, in regard to
each of which there is much uncertainty ; these are, the nature
of the materials of which the portion of the earth beneath the
crust is made up, the amount of condensation effected on this
by pressure of the superincumbent mass, and the reaction of
the internal heat against that pressure.   Of course if the interior of our planet consisted chiefly of metallic iron, or any
other heavy metal, it would have a higher specific gravity than
if silica predominated ; but, even if exclusively formed of a
material as light as quartz, the earth ought, according to physical laws, to be even much denser than it now is, unless there
be some cause acting to diminish the condensing effect of pressure.   Different physicists have made various calculations on
this subject, the results of which are not very satisfactory in
their agreement with each other.   But it is certain that if substances continue to have their density increased by pressure in
descending towards the earth's centre, in the same ratio as they
are found to do at the surface by actual experiment, then we
should have to penetrate to but a few hundred miles in depth, to
find water as dense as platina, and all other substances similarly
compressed.   The force which acts against compression, so as
to make the earth's density, as a whole, only twice that of its
crust, is, in all probability, the expansive action of the internal
heat.   But we know too little of the properties of bodies at
prodigiously high temperatures and under immense pressure,
to say positively whether, under such circumstances, the materials of which the earth is made would have a solid or a liquid
form; and neither astronomy nor mathematics have been able to
give the geologist any valuable assistance in deciding this question.   On the contrary, the most eminent authorities in these
departments of science have published the most contradictory
statements as the results of their investigations in regard to
the condition of the interior of our planet.   Hence, so far as
the astronomical evidence goes, geologists are at liberty to
form their own theories on this subject, and some have inclined
to the belief that the earth is solid throughout ; others stoutly
maintain that it consists of a solid nucleus, with a liquid shell

near the exterior crust; while the prevailing opinion has been that the solid crust encloses a mass of matter fluid nearly or quite to the centre. This latter idea has been naturally adopted, because we are accustomed to see masses of melted metal or stone cool first on the surface, while the interior, if the mass be large, remains for a long time in a fluid condition. The theory of a fluid interior has also been sustained by considerations connected with the widespread distribution of volcanic orifices, and the vast amount of liquid matter which has been poured forth from them at different epochs. The connection of earthquake shocks with the phases of the moon, adverted to in a previous article, is not without an important bearing on this question. The results attained by all seismologists who have investigated these subjects do appear to indicate that there is a decided action of the moon on the interior, analogous to that which it exerts on the waters of the ocean. The evidence is not as decisive as might be wished, but is too important to be overlooked in the discussion of a subject where precise data are hardly to be obtained or expected. The later researches in physics, however, have shown that there is no such sharp line dividing solids from liquids as was formerly supposed to exist; and all the requirements of geology would be satisfied if it should be admitted that the material constituting the interior of the earth, if not already in a liquid condition, was capable of assuming it when relieved of pressure to a certain extent.

All geologists will agree in this, that the disturbances of the earth's crust, by whatever name we please to call them, whenever acting independently of attraction, or against gravity, are due to internal heat. This, in some way or other, is the cause of all earthquake and volcanic action as well as of mountain-building. If the earth were, as the moon appears to be, entirely cooled down, the heat of the sun and the attraction of the sun and moon would then be the sole dynamic agents in producing geological changes. These changes would be affected chiefly through the action of water. The tidal current, raised by the lunar and solar attraction, the powerful, although slowly acting agencies of rain and rivers, — these would be the principal agents of geological change. But these tend almost exclusively to abrade material from the more elevated, and spread it out again upon the lower regions. Hence, the dynamical agencies at work on the earth's surface, supposing the effects of internal heat to be no longer in action, would be directed to reducing inequalities of height; in short, to levelling down the mountains and filling up the valleys. The character of the changes produced by the internal heat of the earth, on the other hand, is antagonistic to this; not exclusively, but nearly so. The proof of this is visible everywhere: in the mountain ranges and single peaks made of lava and volcanic *débris;* in the ranges having an axis of eruptive rock, which has been thrust up from below and carried the overlying stratified rocks with it; and in many other ways.

It being universally admitted that it is the internal heat of the earth which gives rise to the phenomena of volcanism, we have to inquire in what way the results indicated in our previous articles are brought about. The disturbances of the crust by earthquake shocks present the least difficulty in their explanation. Admitting the gradual cooling of the earth as a whole, we find no difficulty in understanding that this cooling may be unequal and irregular in its progress and distribution. This unequal cooling cannot fail to give rise to unequal tension between different parts of the crust; and as the force accumulates until it overcomes the resistance, so, from time to time, as the yielding takes place, there will be a sudden jar or shock given to the surrounding region, which will be more or less severe, according to the amount of force expended in overcoming the resistance. This sort of operation will go on whether the materials of the earth's crust expand or contract on cooling, or even if they, during a portion of the cooling, contract and afterwards expand. That this is the origin of the great earthquakes is proved conclusively by their character and distribution on the earth's surface. Their association with coast lines, mountain chains in process of upheaval, and recent geological formations, affords sufficient evidence that they are not local phenomena, but linked in the closest manner with those other occurrences which have to do with the building up of mountains and the shaping of the outlines of the continents.

Volcanic phenomena, on the other hand, are vastly more

difficult to decipher and refer to their origin, since they are more complicated in every respect, involving chemical as well as mechanical causes and results. To account for all that we know of volcanic rocks is plainly enough a difficult task, since hardly any two eminent authors fully agree in their ideas on this subject. And the larger one's experience and field of observation has been, the more difficult the task of reconciling and correlating all the phenomena has been found to be. Hence, the theories have mostly come from those geologists whose training has been chiefly chemical, and who have looked at nature almost exclusively through the bars of their laboratory windows. Those whose powers of observation have had the widest field for their exercise have had the most vivid perception of the complicated character of the phenomena of volcanic action, and have usually preferred to leave their correlation to others.

The work of Richthofen, the title of which was among those placed at the head of a preceding article on volcanoes, forms an exception to the above remarks, since its author has had an uncommon, in fact almost an unparalleled, range of observation. Having begun with the critical study of the classic volcanic regions of Hungary and Transylvania, he was enabled to carry his researches in an almost unbroken line entirely around the globe, ending with the grandest field anywhere presented to the geologist in this department, the Cordilleras of North America. In several respects this work of Richthofen's, — " The Natural System of the Volcanic Rocks," as it is called, — is one of the greatest importance to the student of dynamical and structural geology. It is the first attempt to go into anything like detail in the investigation of some of the most difficult problems of this branch of the science. That such a work should not meet with immediate attention on the part of the general public was to be expected ; that it should undergo criticism was to be desired, by its author, no doubt, as well as others ; but, that its positive statements of facts of the highest importance in their bearing on the phenomena of volcanism should be overlooked, and even denied, is something which does not argue well for the comprehensiveness or candor of those thus placing themselves in opposition to the introduction of a

" natural system " into that which before had no system at all
connected with it. In the course of this article we shall endeavor to bring out some of the more prominent features of
Richthofen's great paper, and will, in the proper place, give
an idea of some of the criticisms which have been made upon
it. But further light must be thrown on the general subject
of volcanism, before details can be made intelligible to the
general reader.

If we had only the volcanic phenomena of the present day,
or active volcanoes, to deal with, the task of unravelling their
mysteries would, perhaps, not be one of so great difficulty;
but, as soon as we begin to elaborate our materials, and endeavor to correlate the results obtained in the various lines of
research, we find ourselves confronted by an immense mass of
facts going to show that our present volcanic outbursts are only
the last remains, or dying out, of a series of geological events,
the scale of which was formerly much grander than it now is.
We find, without going back to any great distance in geological
history, that there was a time when, instead of being poured
forth from scattered isolated orifices, the eruptive material
found its way to the surface through linear rents, or fissures,
which often must have extended for hundreds, or, perhaps,
even thousands of miles. We find the material which has thus
been poured forth occupying the surface in vast sheets, often
lying in nearly horizontal beds, and covering an area of many
thousand square miles. We find vast chains of mountains almost wholly built up of volcanic rock, and we are able, on careful
examination, to recognize the fact that these masses have not
been brought to the surface in lines radiating from a centre,
that centre being what we call a volcano; but along a linear
axis, in the form of " massive eruptions," as they are called by
Richthofen, who has been the first person to clearly distinguish
between the two kinds of eruptive action, and to give a name
to *massive*, as distinguished from ordinary *volcanic*, eruptions.
The necessity of keeping in mind the difference between volcanic materials erupted from a crateriform opening and those
poured forth from a linear fissure was made evident by Professor Dana more than twenty years ago, in his admirable generalizations on the geological results of the earth's contrac-

tion and the formation of continents.* This idea was also clearly present in the mind of Humboldt at the time of the publication of the first volume of his Kosmos, and Richthofen has in fact carried out some of the suggestions then made by him with regard to the necessity of investigation, by competent lithologists, of the different portions of volcanic ranges which have been piled upon each other at successive epochs and in various ways, And yet we find Mr. Scrope, the veteran author of a much-quoted general work on volcanoes and of the classic description of Central France, denouncing in the most violent language those geologists who think they see any difference in the manner in which volcanic rocks are now and have formerly been erupted.† This fact alone will answer as a sufficient demonstration of the difficulties which the study of volcanic rocks presents, and of the disagreement in theoretical views among geologists, as soon as they begin to enter into details with regard to the mode of volcanic action. ‡

The distinction between massive and volcanic eruptions has been excellently illustrated by Richthofen, as follows: "It is well known that small cones are frequently met with on the slopes of larger volcanoes. If they occur in large number, as on Mount Etna, they are usually situated in lines which radiate from the crater. Each of them is built up of layers of scoria and ashes sloping away from the centre, where a crater is immersed, and such cones will occasionally emit currents of lava, and be in fact the repetition on a small scale of the mother volcano. Just as these parasitic volcanoes have their roots in the glowing lava, volcanoes in general must, as is demonstrated

---

* See "Geology of the Exploring Expedition," and a review of the same in the North American Review, Vol. LXXIV. p. 301, by the author of this article; also American Journal of Science (2) ii. 335 ; iii. 94, 176, 381 ; iv. 88 ; vii. 379.

† See Scrope in Geological Magazine, Vol. VI. p. 512.

‡ Mr. Scrope goes still further in his misconception and misrepresentation of Richthofen's views. He says, "the value of M. Richthofen's " — to an Englishman all foreigners, whether German barons or otherwise, are " M.s " — " Natural System of Volcanic Rocks, as a contribution to the science of geology, may be estimated from the fact that he denies the occurrence of any volcanic rocks in the series of geological formations preceding the tertiary era." The simple fact being that " M. Richthofen " has, for convenience, and following the large majority of authors, chosen to designate the eruptive rocks of the tertiary era as " volcanic," and those of preceding epochs in another manner.

by their mode of occurrence, be considered as parasites on certain subterranean portions of the material of massive eruptions, which still possess a high temperature and are kept in a liquid state by the molecular combination with water which finds access to them."

Richthofen then goes on to show that this mode of origin of volcanoes is only a repetition on a smaller scale of the manner in which massive eruptions themselves originated. Volcanoes bear the same relation to massive eruptions which the latter do to the material forming the primeval interior of the globe. What is this material, and what its relation to the rocks which we call volcanic? These are questions which we have to endeavor to answer.

We must first try to ascertain what volcanic rocks really are. All are familiar with the distinction between igneous and sedimentary rocks, that is, between rocks which have once been in a molten state and which have come to the surface or been deposited through the action of igneous causes, and those which have been deposited by water. Most persons also understand the term " metamorphic " as used by geologists, meaning that the rocks embraced under that term are not what they once were ; that they have suffered certain chemical changes since their deposition, in the course of which the mass has undergone a rearrangement of its particles, so as to have assumed a crystalline texture, separate and distinct minerals segregating out of what was before an amorphous mass in which no particular crystallized minerals could be discerned. Hence, the metamorphic rocks are often called the crystalline rocks. This distinction of rocks into igneous, sedimentary, and metamorphic is, of course, more or less arbitrary. For instance, showers of pumice-stone and ashes may be, and often have been, thrown from a volcano, and the eruptive material falling into water will then have assumed a stratified condition as it sank to the bottom, just as any mud or sand would do. The strata thus formed, having been raised above the water, or while still beneath it, may have undergone chemical changes, or become metamorphic in character, so that the mass now partakes of the character of all three classes.

The formation of sedimentary rocks implies evidently the

pre-existence of some other rock on the earth's surface as the source of the material of which they are formed. Igneous rocks, on the other hand, must have come from beneath the surface, where they have existed from all time, as we may suppose. Believing in common with almost all geologists, that the earth has cooled from a condition of intense ignition, we of course recognize the fact that there was a time when all existing rocks were of igneous formation, — the consolidated crust of the earth was an igneous formation. All the material of the sedimentary rocks must have come from this source; but it may have gone through several cycles of change. Igneous rock has been ground to powder and deposited in water; this material has been again broken up and again deposited ; and no one can say that this process may not in some regions have been repeated a good many times.

It becomes important, then, that some criterion should be established by which the eruptive rocks may be distinguished from the other classes. That knowledge of this kind is needed will be apparent when we consider that the conclusions we have to draw in regard to the dynamical agencies employed in getting the rocks into their present condition and position must depend to a large extent on the origin of those rocks. For instance, if we consider a certain crystalline mass forming the axis of a chain of mountains as an eruptive rock, our conclusions in regard to the structure of that chain will be very different from what they would be if we considered the same material as simply a sedimentary rock which has assumed a crystalline texture from the effects of metamorphic action.

Here, then, we come upon another of the difficulties or discrepancies of opinion among geologists, who, starting from the theory of the original igneous fluidity of the earth, begin almost at once to diverge in their paths towards the goal they wish to attain, which is nothing more nor less than the solution of the great problems of dynamical geology. With regard to the rocks which have come to the surface from beneath during the tertiary epoch, and which we call volcanic, there is but little difference of opinion. We see them now issuing from volcanic vents, and to those products of massive eruptions which precisely resemble in texture and

composition the ejections of existing volcanoes, we do not hesi-
tate to assign a similar origin.   There are, however, many varieties of rocks, occurring in great masses, and belonging to the
older epochs, which were formerly almost universally considered to be eruptive, and in regard to the real nature of which
there is now considerable discussion among geologists.   These
are the rocks of the granitic and porphyritic families.   Granite
and syenite are the predominating types of the granitic, and
quartzose porphyry of the porphyritic family.   These are the
ancient eruptive, or ancient volcanic, rocks in the opinion of
many ; while others look upon them as having been originally
sedimentary, and as having assumed their present crystalline
texture through the influence of chemical changes, — in short,
they are not eruptive, but metamorphic.   By those who adopt
the metamorphic origin of granite and porphyry the argillaceous
slates are supposed to have furnished the material for the first-
named of these, and the sandstones for the other.   If this view
were correct, we should, as advocates of the gradual consolida-
tion of the globe from a condition of igneous fusion, be placed
in a difficult position, for we should have to show how it was
that, in a gradually cooling globe, eruptive material was not
brought to the surface in large quantity until the latest epochs,
when, as would naturally be supposed, the crust of the earth
having become greatly thickened and the interior sensibly
cooled, eruptive action would have diminished instead of having increased.   To avoid this difficulty, some of the chemical
geologists — and of those who maintain the metamorphic origin
of granite such are indeed the only consistent ones — deny
altogether the existence of any primeval eruptive rock.   To
them all visible rocks are either sedimentary, or they have been
such ; and what are ordinarily called volcanic and eruptive
masses are nothing but sedimentary deposits which have been
softened or liquefied by the internal heat, and thus enabled to
flow as lava.   The idea of these geologists seems to be, that
the series of changes has been going on so long on the earth's
surface that no portion of the original crust can, by any possibility, remain visible.   It is a pushing to its extremest limits of
the favorite theory of Lyell, that no traces of a beginning are
to be found ; or, at least, that the beginning is to be put back

as far as possible, and that all geological phenomena are to be interpreted with that one idea in view, the result being that some facts have been extraordinarily distorted and others overlooked, for the purpose of making things pleasant in that direction. Such persons as wish to make it appear that no proofs of a beginning can be found in geological facts must go still further, and deny that the earth has ever been in a condition of igneous fluidity, from which it has been gradually cooling through all the geological ages. They are trying to pull out the corner-stone from under the fabric of the science.

It can be clearly shown, as it seems to us, not only that the volcanic rocks are not softened or metamorphosed sedimentary materials, but that the same is true of the rocks of the granitic and porphyritic families; these are, in fact, samples of the primeval crust of the earth, such as it was before any sedimentary rocks had been formed. In order to get at some of the proofs of this, it will be necessary to consider, for a moment, the mineralogical composition of the different families of the eruptive rocks; these are all almost exclusively aggregates of silicious minerals, including among them silica itself or quartz. Several different kinds of feldspars; hornblende and augite, two very closely allied minerals; quartz; different varieties of mica; magnetic iron: — these are the substances of which all eruptive rocks, including granite, porphyry, and lava, are almost exclusively made up. Quite a number of other minerals do indeed occur in them, but almost always in very subordinate quantity. The close resemblance in external appearance and actual composition between eruptive rocks from different parts of the world is, indeed, a surprising fact. But it is more surprising still to find that, as shown by the researches of the great chemist Bunsen, the materials of which these rocks are made up are combined in certain definite proportions; so that if we determine by chemical analysis the quantity of any one of the ingredients of which a specimen is composed, we can by mathematical calculation arrive very nearly at the amount of each of the others. The "law of Bunsen," as it is called, is of the greatest possible importance in its bearing on the question of the origin of the eruptive rocks. It must be evident to all that this law could not be true if the rocks to which it applies were of metamorphic origin. If that were the case, and they were really derived from the sedimentary deposits, they could not, by any possibility, fail to have the same varying composition which these sediments themselves have, and which can by no means be brought under Bunsen's law.

There is also another fact which has a most important bearing in this connection. It is this : that the order of succession of the volcanic rocks has been the same all over the world ; they have not come to the surface in different regions in an indiscriminate manner, but in a certain sequence, or chronological order. This extremely important fact was first brought out by Richthofen, who, by means of his specially good opportunities for the study of this class of rocks, was enabled to recognize and clearly lay down this order of succession, and demonstrate its correctness by examples collected all over the globe. The chronological order of succession, as well as the law of composition of the volcanic rocks, are clearly opposed to the idea that these are the results of the metamorphism of the sedimentary beds. The material of which these volcanic ejections are made up must have come from beneath the shell of sedimentary deposits ; and as it everywhere came from beneath this shell in a certain chronological order, so it must ever have previously existed there in the same order. If basalt has always been erupted after andesitic lava, then basalt must have everywhere formed a shell of material underlying andesite in the earth's interior ; that is to say, the mass of the earth beneath the shell of sediments is formed, for a certain distance down, of layers of somewhat different material, and these layers are arranged in a similar order all over the world. What is this order ? Is it one in which we can find something logical, something which seems to be connected with the nature of the materials themselves ? To this question the answer is, in a general way, affirmative ; but it must be admitted that the processes of volcanism are so complicated that we cannot expect an agreement in all minute details, but only in the general order of events, looking at them in the largest way. It will not do to study up the exceptions to the general rule and make them our standards, as we are likely to do if we confine our observations to any one locality. We are rather to try and get the general principles established, and then endeavor to account for the apparent exceptions in a manner which will be in harmony with the general well-established series of facts. Thus, if it can be shown that over nine tenths of the globe the order of succession of the volcanic rocks is one and the same, then let this fact first be thoroughly demonstrated, and afterwards let the exceptional cases in the remaining tenth be investigated, each on its own merits, in its necessary subordination to the general law.

In something like this spirit the investigations of Richthofen, in regard to the order of succession of the volcanic rocks, must be received. It is not claimed that he has clearly made out their precise sequence in all localities and for all geological epochs ; but that there is a certain order to which they have conformed, over a large portion of the earth, and especially during the tertiary period, can no longer be doubted ; while it seems probable that the exceptions which do occur will be found to be of comparatively slight importance, and that all geologists will have to admit the value of these investigations in their bearing on the difficult questions to which they are applicable.

It certainly seems clear enough that, on the whole, the order in which the volcanic rocks have appeared is one which we ought to have expected, if the theory of a gradually cooling globe be true. The more silicious and, of course, the lighter kinds were the first to be emitted from the interior, and these have been succeeded by denser or more basic ones. This statement is not so peculiarly applicable to the volcanic as it is to all eruptive rocks, beginning with the earliest epochs and including the granitic family. From this point of view it is evident that quartzose and the more highly silicious rocks prevailed almost exclusively during the earlier periods, and that they have gradually become replaced by the more basic. Granite and syenite were once the predominating eruptive rocks ; in the latest geological ages basalt and andesite have been.

As the development of the earth's history has gone on, the regions of igneous action have become more and more localized, and we have now only eruptive materials issuing from craters

7

or isolated orifices; the days of massive eruptions, or such as took place from fissures of great length, have passed. This is as we should expect; for, although there are some who follow the school of Lyell so far as to reject everything which looks like more violent action of any kind in the past than at present, yet, unless we admit that igneous forces were more actively at work and more generally disseminated than they now are, we must give up altogether the hypothesis of a gradually cooling globe; and, with this theory gone, we are entirely afloat,—absolutely destitute of any guide through the mazes of structural geology. We must admit that the crust has been constantly thickening, while the cooling has been going on; and if this has been the case, the facility with which the molten matter in the interior has found its way to the surface must have been constantly diminishing.

There is a point, in this connection, to which our attention must be for a moment turned. There is a difference between the granitic and the ordinary volcanic rocks, as regards the method in which they have come up from beneath, dependent on their position as portions of the exterior shell of the earth, in consequence of which the former have more of an intrusive, and the latter rather an eruptive character. Forming, as it did, the original surface or uppermost layer, granite has often been raised in ridges, before any sedimentary rocks existed, through which it must otherwise have been obliged to force its way. There being no resistance from the weight of overlying materials to be overcome, this rock could assume a higher position without having to wait for tension to accumulate so as to form fissures, as has been the case with the more recent eruptive masses, which have had to find or make in some manner a passage through a considerable thickness of the consolidated crust, before they could appear upon the surface.

The real character of granite and the granitic rocks has been much discussed of late among chemists and geologists, the former adopting usually the metamorphic theory of its origin, the latter, on the other hand, almost all taking the other side. The field geologist sees these rocks occupying a position which it seems impossible that they should have, unless they have been forced upward when in a liquid or plastic condition, and he observes also a great many facts which preclude the idea that this liquidity or plasticity has originated through metamorphic action on sedimentary materials. And in a question of this kind, at present the geological facts must be allowed a greater weight than the chemical, since chemistry has thus far proved to be rather a blind guide to those endeavoring to unriddle the mysterious reactions of the primeval earth. That the peculiar texture of granite, as compared with that of the volcanic rocks proper, does present a difficulty, there is no doubt; but if we consider that this rock, forming as it did the exterior crust, must have been in much closer proximity to the ocean than were the underlying masses, we shall have no difficulty in understanding that a larger amount of water and a lower temperature were conditions which exercised a powerful influence in determining its texture.

There can be no doubt, then, that the seat of volcanic action has gradually receded from the exterior towards the centre, and that in so receding it has descended into regions of denser material, and that these regions have been reached in the same order in different parts of the world, showing that the arrangement of the materials of the crust is everywhere strikingly similar.

We have an important and difficult question to answer in endeavoring to ascertain the nature of the force which brings the material of the molten interior to the surface. This is a subject which has been passed over without discussion by some writers, while others have given it a measure of consideration, usually making it evident by their treatment of it that they felt its difficulties. It used formerly to be supposed that the opening of a fissure in the earth's crust would necessarily cause the molten material below to issue forth without further cause. The insufficiency of this as a reason has been felt by the later writers; but of those discussing the subject hardly any two have been agreed in their views. Sometimes the differences of opinions thus disclosed are not radical; but usually they are, and in a good many instances we find authors diametrically opposed to each other. In one respect there is a fair amount of agreement among the theorizers on volcanic phenomena. Almost all consider the access of water as essential, in some

way or other, to the emission of lava. But in regard to the
modus operandi of the water and the manner in which it is to
find its way down to the volcanic focus, most authors are found
to preserve a discreet silence. Mr. David Forbes, whose lec-
tures and writings on these subjects have been much circulated
of late in the English magazines, says that all which is required
to account for the phenomena of volcanic action is "the assumption that water from the sea should, *by some means or other*, find
its way down into the reservoir of molten matter beneath the
surface"; what the means are by which the water is to gain
access to the interior are not given, nor is the mode in which
the water acts after it has reached the depths anywhere explained. Scrope, who was among the first of modern authors
to advocate the necessity of water as an agent in volcanic eruptions, solves the difficulty in a most curious manner, namely,
by supposing the water to be already present in the material
which is to issue forth as lava, and only waiting to be vaporized
whenever a transfer of heat into the region takes place. To use
the words of that author, "It is now generally recognized that
the power which forces up lava from a depth of miles, through
narrow and crooked fissures broken across the solid crust of
the globe is no other than steam, developed in the interior of the
lava by vaporization of water intimately disseminated throughout its substance." Professor Phillips, one of the most cautious of the English writers on geology, in his latest work on
Vesuvius, quoted in a previous article, incidentally alludes to
water as a cause of "volcanic excitement," as he terms it, but
goes no further in that direction.

The eminent chemical geologist, Bischof, is the only author
who has gone into anything like an elaborate discussion of the
manner in which water might gain access to the molten interior and act as a motive-power in the ejection of the lava. He
perceives some of the difficulties in the way of the adoption of
this idea, and endeavors to remove them. It seems pretty
clear that steam at its maximum elastic force would not have
power enough to raise a column of lava from the region from
which it is supposed to come up to the summit of even a mod-
erately high volcano. This difficulty Bischof gets over by supposing that the column of lava has lengths of steam included

in it, like the bubbles of air in a barometer tube.   This explanation is also adopted by Lyell, who follows Bischof closely in all that relates to the theory of volcanic action.   This hypothesis, moreover, clearly involves another difficulty, which is this : that two columns, one of water and the other of lava, must be in communication with the molten mass of the interior of the earth, and yet that the elastic force of the steam generated by that water shall throw out, not the water itself, but the lava.   It is believed by some physicists that it may be possible for water to pass through minute fissures, through which it cannot return when converted into steam, although this has not yet been clearly demonstrated.   But, even admitting this, it does not appear how it is that the force of the steam is used to lift up the column of lava to a height of ten, fifteen, or even twenty thousand feet above the level at which the water enters, rather than to blow out the fissured and necessarily much weakened thinner portion of the crust through which the water has found its way.   This objection is an insurmountable one, in our judgment ; and, indeed, the assumption that steam is the *primum mobile* in all volcanic eruptions is one beset with difficulties. No theory of volcanoes can be adopted that will not account for the phenomena of massive eruptions as well as for ejections from crateriform orifices, and this the water theory is obviously incompetent to do.   That water comes into play in volcanic eruptions there can be little doubt; but this is in the later stages of the process, when cinders and ashy materials are chiefly ejected.   And it is by no means certain that the rain may not be quite as competent as the sea to supply the necessary water.   Much stress has been laid on the fact that most volcanoes are near the sea or on islands, as going to prove that eruptive action cannot take place without the presence of seawater.   But it must be recollected that this nearness is, in many cases, only comparative, with reference to the total breadth of the continent, and not absolute.   Thus the volcanoes of the South American Andes are, in many instances, two hundred to three hundred miles from the sea, which is certainly a long distance for action to be transmitted laterally through the intervening rock.   Besides, there are not a few regions where, within a recent geological period, if not during the present epoch, volcanic action has taken place on a large scale, at a great distance from the sea, at a high altitude above it, and also far from any inland waters of magnitude, which might be supposed to answer instead of the ocean as feeders to the volcanic excitement. We need only instance, in this connection, the line of volcanoes which extends across our continent through Northern Arizona and New Mexico, of which Mount San Francisco and Mount Taylor are the dominating summits.

Everything indicates that we cannot separate the agencies which give rise to the formation of mountain chains from those which are energetic in volcanic eruptions. Whatever cause is capable of folding the crust of the earth into ridges, or thrusting a portion of it up above the adjacent parts, is also competent, if carried a little further, to produce a fissure, and through this the underlying material, whether it be in a fluid, plastic, or viscous condition, may be forced, by the pressure arising from the subsidence of that portion of the crust which borders it on one side or the other.

We have, therefore, to go back another step and endeavor to ascertain what the forces are which have been active in producing those ridges of the earth's surface which we call mountains. A mountain may result either from a positive elevation of the mass, or from depression of the adjacent region. We leave out of view here those elevations which have their origin simply in denudation or erosion by water of the surrounding surface, for these are easily understood and comparatively unimportant. It is true that we have absolutely no means of ascertaining how much, in the past geological ages, of the elevation of our mountain chains is due to actual upheaval or increase of distance from the centre of the earth, and how much to depression of other portions of the surface. We are accustomed to refer all elevations to the level of the sea as a zero, but we have no reason to suppose that this level has itself been invariable ; that is to say, it cannot be taken for granted that the distance from the centre of the earth to the sea-level at any particular point on the earth's surface has always remained the same. On the contrary, there is abundant evidence that the sea-basins have deepened since the earlier geological periods ; but of the extent to which the sinking of the sea-level which would thus be produced has been compensated by an increase of the area of the land, we can only form the crudest conjecture. We do not yet know the depth of the deepest portions of the ocean, or where they are situated.

Looking at the surface of the earth simply with reference to continental and oceanic areas, we have reason to believe that the differences of level between them are the result of depression rather than of elevation. The masses which now form the continents have been left where they were, while the ocean beds have sunk and allowed the water to retire from the more elevated portions. This follows, indeed, necessarily, from the nature of the assumed cause of differences of elevation, namely, the shrinkage of the interior, and the endeavors of the crust to adapt itself to the diminished nucleus. If we conceive that the globe, as a whole, shrinks somewhat unevenly, and it is hardly possible to conceive that it should be otherwise, since neither the composition of the cooling body nor its rate of parting with its heat would be likely to be entirely uniform in all its parts, then the region in which positive elevations would be likely to take place would be the borders of the most rapidly shrinking area, or where it joins on to the portion which remains comparatively stationary. These more rapidly shrinking areas would, of course, be the ocean beds, and the stationary area the conti nental masses, while the edges of the continents would be the region of positive uplift or of mountain formation. This is the basis of Professor Dana's theory of the formation of continents, as set forth by him in the " Geology of the Exploring Expedition " and elsewhere,* and of which a synopsis was given by us in this Review some twenty years ago. The investigations of geologists have, since that time, given additional value and lustre to these lofty generalizations of Professor Dana's, in the opinion of the writer of this article, although it must be admitted that they have not met with general adoption. Other theories have been suggested and discussed, but without any very definite conclusions having been arrived it; at all events, the conclusions reached have rarely been satisfactory to others than their authors. A great mass of material has been gathered bearing on the structure of mountain chains, or, at least,

---

* See references on page 79.

capable of being made available in that direction; but little has been accomplished in the way of applying this information to the working out of any theory of mountain building. What theories have been suggested have been of the vaguest kind, and, in some instances, facts have been entirely ignored in supporting them.

While believing, with Professor Dana, that mountain-building is, to a large extent at least, the result of an antagonism between subsiding and stationary masses of the earth's crust, we are fully aware this is a somewhat vague way of stating the case, and that a more detailed account of the agencies at work in this operation, and of the methods in which they act, is extremely desirable. But when we come to examine what is known of the detailed structure of the great mountain chains of the world, we find that, in spite of all that geologists have done, our information is exceedingly defective. In the chain of the Alps it is true that we have a great many local sections in the works of Gümbel, Favre, Studer, and especially of the geologists of the Austrian official survey, the Reichsanstalt. But how deficient are our generalized sections across the entire chain! Indeed, there is not one on a large scale from which an idea of the structure of the mass, as a whole, can be obtained. And if this be true for the Alps, how much more is it likely to be so for the great chains of Asia and of America, which, in comparison with the much visited and studied European mountain masses, are almost unknown. Indeed, it is only quite recently that the subdivisions of our own Cordilleras, grand as they are, began to be indicated on our maps or even to receive names. And not even so far as that has our knowledge of the Asiatic chains of mountains reached. Of that vast region north of the main Himalayan range, on which are piled the masses of the Kün-Lün, the Karakorum, and others, we know as yet almost nothing, so far as geological structure is concerned. Even the sections which the India survey gives of the middle and lower Himalayan ranges are on a small scale and difficult to unriddle. Generalizations in regard to mountain structure at the present time, which profess to go into some detail, must, therefore, be drawn with much caution, and taken rather as indicating the direction in which future and much-needed work may be accomplished, and not as based on anything completed.

Of all mountain forms, the simplest are those which result from denudation. Masses of rock are often left standing, isolated from each other by the removal of the adjacent material through the action of water; and these masses, where the erosion has been extensive and long continued and in suitable strata, are occasionally so large as properly to be called mountains. There are fine examples of the forms resulting from erosion in our Rocky Mountain region and farther west. But erosion on a large scale cannot take place without continental elevation. There must be a rapid inclination of the surface towards the sea to admit of portions of the surface being deeply cut into by the streams which traverse it. Hence the formation of mountains by erosion is rather to be regarded as a secondary operation, and as a sort of carving of an already elevated mass into detached portions which may then bear the name of mountains, which, previous to the erosion, the whole would have been called simply a plateau. The Book Mountains in Colorado are admirable instances, on a grand scale, of this mode of formation.

The next most simple form of mountain building is that in which masses of rock — and it is chiefly the sedimentary formations which are thus acted on — are broken across and tilted up at an angle, from an unequal subsidence of the fractured portions; something as we see happening in the ice covering the surface of a lake when it has been broken up by the waves and then frozen together again, the different pieces being inclined to each other at slight angles, instead of lying all in one plane as before. Such mountains are not usually developed on a large scale, for in almost every case, if there is a fissure formed, there is an outpouring of eruptive material.

If, on the other hand, the group or series of strata, instead of being broken across and tilted, are gradually bent, then a ridge or protuberance of the surface will be formed, and it will have, of course, various degrees of curvature. A series of such ridges will alternate with relatively depressed regions or valleys, the whole forming a system of foldings which are very likely to be parallel or nearly so, because parallelism in this case merely means a persistence of the bending agencies in one direction. Such a system of parallel ridges, or folds may be seen in the Appalachians and the Jura, two perfectly typical regions in this respect. But these may not, by any means, be taken as representatives of all mountain chains, as has been done by Hall and H. D. Rogers. On the contrary, they are only chains of the second or third order of magnitude, so far as elevation is concerned, and in many respects exceptional. They are, so far as we know, the only systems of mountains, having great geographical development, in which there has been no emission of eruptive material from below and no extensive metamorphism.

It seems to be clearly indicated by the results of geological investigations, that the great mountain chains of the world have been blocked out — if the use of such a phrase may be permitted — from the earlier geological times, and often from the earliest. Their structure shows most distinctly that their development has been a gradual one. But it was not always the case that this development was continued down to the latest period. On the contrary, many chains have ceased to grow after attaining a certain elevation ; and, having ceased to be influenced by forces acting from beneath, they have ever since been subjected to those erosive agencies which constantly tend to plane down the inequalities of the surface. Hence, the highest chains contain the most recent geological formations. The Himalaya, the Alps, the Andes, the Cordilleras, — these are the great chains of the world, and these are all made up, in part at least, of the newest formations. The Ural, the Scandinavian Mountains, the Appalachians, the Brazilian ranges, — these are examples of mountain chains which have ceased to grow at a comparatively early geological period, and within whose masses no modern rocks can be found.

The results of modern investigations, especially in the Andes and Cordilleras, are diametrically opposed to the theories of Elie de Beaumont, on which he has spent so much labor, and which he has built up with such care and such an outlay of mathematical calculations. According to the views of this eminent French geologist, the earth in cooling and contracting has developed its mountain ranges along lines which are parts of great circles.drawn about the globe in a network of curves developed symmetrically from the points where a solid with regular pentagonal faces included within the earth would touch its surface. It is also a part of De Beaumont's system, that mountain chains having the same direction must be of the same geological age; so that law, order, and crystalline harmony would seem to be clearly established in what would otherwise seem almost a chaos of facts, if these theories should bear the test of close examinations, and found to be applicable all over the globe. So desirable was this, that it is no wonder that many geologists were glad to become converts to these views. One by one they have dropped off, however; and few excepting Frenchmen are now found upholding the theory of the pentagonal network. Many years of labor among mountains of the first rank have convinced us that the real facts are almost exactly in opposition to Elie de Beaumont's views. Instead of its being true that identity of direction in mountain chains implies identity of geological age, one might almost say that just the opposite is true. Certain it is, that the great mountain chains are made up of distinct portions, which have similar directions and very different geological ages. Thus, in the Andes and Cordilleras, we have one grand system of mountains made up of an aggregation of many different parts, each having approximately the same direction, and each of these parts or sections being the result of a series of geological changes which have been going on through all the epochs, from the earliest to the latest.

Take, for instance, the widest portion of the whole belt of mountains which forms the western side of the American continent, or that between the thirty-sixth and fortieth parallels of north latitude. We have here a mass of ranges fully a thousand miles in width, having a certain unity which cannot be disputed, and yet made up of parts which have been growing on to each other ever since the azoic period. For, even at that earliest geological epoch, the chain of the Rocky Mountains was marked out, and each successive period, down even to the very latest, has seen some additions made to the mass.

In all great and complicated chains of mountains, almost without exception, we find eruptive rocks forming a portion of the mass; these may be either ancient or modern, or both to-

gether. Great chains almost invariably are made up, to a large
extent, of granitic rocks; usually granite itself forms the bulk
of the mass. Volcanic overflows may or may not occur; different chains differ very much in this respect. The granite
usually forms the central and higher portion of a great chain;
it is a remarkable exception when this is not the case. In the
Alps, while the bulk of the central masses are of a granitic
character, there are a few very lofty and almost isolated points
or even large domes made up of sedimentary materials, as, for
instance, the Matterhorn, and some of the very highest portions
of the Bernese Oberland. In the Himalayas the main portions of
the higher ranges seem to be granite, but data are extremely
deficient for those regions. Eruptive rocks, both of the granitic and volcanic types, are abundantly but very unequally disseminated through the great ranges which make up the Pacific
edge of North and South America. The Andes are very largely
made up of volcanic materials piled on each other to an immense height; these appear to predominate over the granitic;
but different portions of the chain are very unequally situated
in this respect. The same is true with regard to the North
American Cordilleras; here, vast masses of granitic rocks
forming exclusively all the more elevated ranges; there, volcanic materials covering up all the others, and far exceeding
them in quantity.

When the study of orography was in its infancy, it was
thought that the typical form of mountain ranges was that
of a mass or wedge of granite thrust up from beneath and
carrying with it the sedimentary rocks through which it had
made its way, which would then be symmetrically disposed
upon the central mass, the stratified beds dipping each way
from it, and forming what geologists call an anticlinal axis.
It was found, after more accurate observations began to be made
than were customary in the early days of geology, that the
structure of most of the great chains was by no means so simple as this, and, consequently, some hastened to conclude and
to state that no such thing had ever occurred at all. Some
even went to such an extreme in the opposite direction as to
maintain that all mountains had a structure exactly the reverse
of the anticlinal, namely, synclinal. Of this theory more pres-

ently ; it must be considered in connection with that which makes granite and all the granitic rocks to be of sedimentary origin, and not eruptive, but metamorphic.

Believing, as we do, that granite or some rocks of the granitic family formed the original exterior crust of the earth, it is not difficult for us to understand that these must necessarily form the core of most mountain chains, and that especially it must predominate in those which reached their full development during the earlier geological ages. When the ridging or wrinkling of the crust began to take place, granite, being the uppermost layer, was raised into the highest position, and might be elevated to almost any amount, provided the base on which the protuberance was raised was broad enough. Circumstances, the exact nature of which it would not, in the present state of our knowledge, be easy to state in detail, have differently influenced the different ranges in regard to the point whether the granite crust should be entirely broken through and the underlying more basic rocks be brought to the surface. In the Andes and Cordilleras, everywhere the eruption of the granite has been followed, at some stage of the mountain-building process, by the outpouring of volcanic rocks, beginning with propylite and andesite and ending with basalt. We know too little of the structure of the great South American chain as yet ; but it is certain that modern volcanic rocks form a large portion of it, and that granite lies at the bottom of the whole, although subordinate in quantity, at least through considerable portions of the chain. In North America the granite predominates, on the other hand, and the volcanic, although crowning the range in many places, is, on the whole, much inferior in bulk to the more ancient eruptive masses. This relation is changed, however, as we go north, and in Oregon basaltic lava covers almost the whole of the Cascade Range, and has flowed far and wide over the adjacent country. Striking as is the predominance of volcanic rocks in the mountain ranges which encircle the Pacific, it is still more extraordinary to find them almost wholly absent in the High Alps, and in the Himalayas, so far as yet ascertained, while abundantly exhibited both north and south of these ranges. Thus the vast lava plains of the Dekkan lie to the south of the Himalayas, while to the north extensive volcanic formations are also reported; but so little is accurately known of that region, that it is hardly possible to say whether there are any traces of active volcanism there. The volcanic formations of Europe lie to the north and south of the Alps, at a considerable distance, as any one may see by consulting a geological map of that country. The best solution which can be offered for this problem of the unequal distribution of volcanic rocks on the two opposite continental masses is, that in Europe-Asia the thickness of the granitic crust was greater than on the American side, so that the underlying volcanic masses could not find their way to the surface through the uplifted protuberance, but only at its edges, where tension was great and the thickness of the granitic layer less than towards the centre of the uplift. That this may have been the case is indicated by the much greater extent of the land mass of the continent of Europe-Asia, the greater absolute height, and the vastly greater breadth of the ranges taken as a whole. When these die out, then the volcanic rocks come in, as to the south of the Caucasus and in the space between that chain and the western extremity of the Himalayan ranges. It is not without a meaning in this connection that, as it appears, the phenomena of absolute elevation have been continued up to a later geological period in the chains bordering the Pacific than in that region which includes the Alps and the Himalayas.

The mechanism by means of which simple upheavals, uplifts, or downthrows of portions of the stratified shell of the globe are accomplished is not difficult to be comprehended. But, to explain the origin of so complicated a series of folds as that exhibited by the Jura and the Appalachians, where there is no central axis of crystalline or eruptive rock, is a more difficult task. Among the theories proposed to that end, that of Professor H. D. Rogers is the wildest and most fantastic. According to this, it was the pulsation of earthquake waves through the molten interior of the earth which laid the superficial crust in plaits. As this idea has never met with acceptance on the part of any sober-minded worker in geology, it need only be alluded to here. If it had not been elaborated with so much care and brought forward on so many occasions by its author, it would have seemed as if rather intended to be classed with that half-playful hint of Sir John Herschel's, that the heat of the sun is kept up by monstrous organized existences, whose dim outlines are revealed to us in the willow-leaf structure of the surface of our "ruler, fire, light, and life," as Mr. Proctor calls the centre of our planetary system. There is nothing about Professor Rogers's theory which will bear the test of examination. It has not the slightest adaptation to chains which are unlike the Appalachians in structure, and, as already stated, this range and the Jura are quite exceptional in character. From it we get no clue as to how the waves originated; how they were propagated from one side only, as would be required to meet the case of the structure of the Appalachians; how the strata, instead of being shattered in pieces by the rapid pulsations of the internal fluid, were gradually bent in such a manner as could only have been accomplished by very long-continued action; how the corrugated crust was held in place after the passage of the wave. In fact, from whatever side we examine this theory, it presents nothing but difficulties, of which only a few have here been suggested.

Another theory of mountain formation, which was first intended to be applied to the Appalachian chain, but which has since been stretched to fit all mountain ranges, is that of Professor James Hall, which has also been supported by Mr. Sterry Hunt, and by Mr. Vose, in a work entitled "Orographic Geology." This last-named gentleman, who prints "civil engineer" after his name on the title-page of his work, as if he feared that, by some possibility, he should be taken for a geologist, has adopted Mr. Hall's theories *in toto*, which he could more easily do, since he was not hampered by any of those difficulties which have their origin in a personal acquaintance with the subject.

Professor Hall's theory is rather an application or enlargement of the views of Herschel and Babbage in regard to the manner in which the internal heat of the earth may be supposed to affect regions where deposition or denudation of the strata are taking place. As it is known from observation that the isogeothermal lines, as they are called, that is, the lines of equal temperature beneath the surface of the earth, rise and fall with the elevations and depressions of the surface, so that

the underground isothermal surfaces correspond in contour
with the external surface. This being the case, if over a certain region there is a deposition of sediment going on, then
there must be a rising of the temperature beneath while the
isogeothermals are adapting themselves to the new surface.
Exactly the opposite will take place in a region from which the
material is being abraded. Thus, as erosion and deposition of
sediments are always going on, there are always changes of temperature taking place over the earth's surface, by which expansion and contraction of the rocks are effected. This is pre-
sumed by Babbage to be an agency of the first importance in
producing geological changes, and Herschel also insists upon
the increase and relief of pressure in different regions, according as material is deposited or abraded, as also necessarily
being one of the mightiest of the causes by which changes in
the configuration of the surface are brought about.

These views have been applied by Professor Hall in this way.
Deposition of sedimentary materials can only take place con-
tinuously and for a long time in a region which is subsiding, as
all geologists will readily admit, since detritus must be carried from a higher to a lower region, and if that less elevated
area does not subside it will soon be filled up with sediment.
Subsidence, however, according to Professor Hall, involves plication or folding of the strata, which must take place when
large thicknesses of material are pressed downwards. To use
the Professor's own words : " By this process of subsidence, as
the lower side becomes gradually curved, there must follow, as
a consequence, rents and fractures upon that side ; or the diminished width of surface above, caused by this curving below,
will produce wrinkles and foldings of the strata." Further on
he adds : " But the folding of the strata seems to me a very
natural and inevitable consequence of the process of subsidence." The results are, according to this theory, that mountain chains do not occur except where there is a great thickness of sedimentary deposits, and that these become plicated by
their own subsidence. Hence plication is characteristic of all
mountain chains ; so, also, is metamorphism, for in the subsidence the material has been brought into such relations of
position as to cause the isothermal planes to ascend into it,

and thus to bring it into such conditions of temperature as to facilitate those chemical changes which result in converting a sedimentary into a metamorphic rock. Hence, also, a synclinal structure and an axis of metamorphic rocks are to be expected in every great mountain chain. But how the mountain chain is obtained from the depressed mass of strata is nowhere explained by the author of the theory in question; hence it has been aptly characterized by Professor Dana as " a theory for the origin of mountains, with the origin of mountains left out." Indeed, there is no point in which it will stand the test of examination. It admits of mathematical demonstration that the assigned cause would not be sufficient to cause the plication. This can also be made apparent to the eye by drawing a diagram representing a section of a portion of the earth's crust on a natural scale, laying off an area of subsidence with an amount of depression equivalent to the assumed thickness of the stratified rocks, say of the Appalachian chain, and observing the relative length of the lines representing the original surface and that of the depressed mass. The result will be quite conclusive as to the plication of strata from their own subsidence, except where that subsidence is extremely local. Neither is it true that mountain ranges exhibit usually anything like the kind of synclinal structure required by Professor Hall's theory ; indeed, if we can understand what this structure would be most likely to be, there is no such chain anywhere. The theory, as set forth by its author, is left in such a vague form that it seems impossible to bring it to any crucial test, and one has to be content with finding in it nothing which will bear examination.

It must be borne in mind also, in this connection, that neither Babbage nor Herschel were geologists, and that, consequently, their views with regard to the relative importance of different geological agents or conditions are not to be accepted without careful investigation. A little consideration will show that although there may be something plausible, and even attractive, about these theories of metamorphism and change of relief of the surface in consequence of denudation and the accumulation of sediments, the facts are far from supporting them, at least to anything like the extent assumed by Professor Hall. If the earth's crust is so sensitive to pressure that it is

8

ready to respond to the very gradual and comparatively slight difference of level resulting from abrasion of the rock at one locality and removal of the detritus thus formed to another, how is it that the weight of the great mountain masses is supported, or how could they have originated at all? It is not possible to conceive that, during all the preceding stages of the earth's existence, its interior should be so insensible to the pressure of the crust as to allow ranges like the Alps, the Andes, and the Himalayas to be built upon it, and that, at the present epoch, it has, all at once, assumed such a condition of sensitiveness as to respond by its motion to any transference of weight from one region to another. It would not be difficult to suggest other valid reasons for refusing to accept Herschel's views; but enough has been said to indicate clearly that they are not admissible as a basis for orographic generalizations. The ideas of Babbage in regard to the rise of the isogeothermal planes in consequence of the accumulation of sediments are more philosophical than those of Herschel; but the facts do not bear us out in inferring that extensive metamorphism will necessarily be the result of the resulting increase of temperature. An examination of a section of stratified rocks piled upon each other to a height of several thousand feet, resting horizontally on the granite, and quite unaltered as to texture since deposition, is sufficient evidence that heavy accumulations of sediment are not necessarily rendered crystalline by the rise of the isogeothermal planes; but that something else is required to bring about that complex series of chemical changes which we designate by the term "metamorphic action." Such sections as those alluded to here may be seen in abundance over a wide area in the Rocky Mountains, and along the Colorado and its tributaries, as well as elsewhere. Indeed, the Appalachians and the Juras themselves show that great masses of rock may be piled up, and even extensively plicated, with but little resulting metamorphism.

The gist of Professor Hall's theory seems to be, that mountains are logically connected with large deposits of sedimentary rocks; and this is true, but exactly in the opposite way from that imagined by him. The sedimentary beds are thick because the mountains pre-existed from the destruction of which they could be formed ; not that, having been already formed, they were afterwards made into mountains. There can be no formation of detrital or sedimentary deposits, that is, of stratified rocks, without the previous existence of some higher region from which the material can be derived. Hence, if the combined thickness of the sedimentary beds about a great mountain centre reaches a high figure, it is simply because the conditions for the accumulation of such beds have been favorable. With a surface entirely flat, the amount of deposition must necessarily be very small and almost entirely confined to such materials as are produced by chemical or organic action. But those beds which are chemically precipitated or formed by living organisms are vastly inferior in thickness to those which result from the piling up of detrital materials, or such as are abraded from previously existing rocks through the agency of water.

It is evident that, in theorizing in regard to mountain-making and deposition of sediments, too little regard has been had to the origin of these sediments. The fact is ignored that all the sedimentary formations must have been originally derived from the original crust of the earth as it existed after cooling had gone so far that water had begun to condense upon its surface ; they must have had some higher region from which to be swept downwards. These higher regions were, in the first place, evidently the ridges or wrinkles of the granitic and gneissoid crust raised above the general level by the first efforts of the consolidated crust to adapt itself to the interior. The detritus thus carried down the flanks of the ridges was, early in the geological history of the earth, mostly deposited in the ocean, which must originally have covered even a larger portion of our surface than it now does. Hence the predominance, or almost exclusive existence, of marine formations, during the earlier geological ages. It was not until a large body of sedimentary deposits had thus been formed, and these masses had begun to be themselves raised above the sea-level, that their abrasion could furnish material for a set of beds not derived from the original crust. And this process having once been gone through, the same thing may have been repeated again and again. How many times such a destruction of pre-existing sediments and formations of new deposits from the ruins may have taken place in any one region, we cannot say; but we have no reason for assuming that all over the world this has gone on to such an extent that none of the original crust can be anywhere visible.

The area of the continental masses gradually and constantly expanding, and the depth of the oceanic basins increasing, strata formed by fluviatile action began to be deposited, and of course contained the remains of fresh-water and land animals. If, then, no new axis of elevation was originated, and there was no further rise of the land, the formation of new stratified deposits would eventually reach its limit, because the newly formed beds would have risen to the level of the highest existing land, and, equilibrium of the surface having been restored, there could be no more erosion, except on the smallest scale. Thus, in many mountain chains, as already noticed, there has been a cessation of growth at an early period; while in others — and these are the great chains of the world — growth has continued down even to the very latest epoch. In these instances of continued growth there has usually been a tendency to the formation of a new axis or uplift parallel with the earlier one, and at no great distance from it, on one side or the other. Thus opportunity has been given for the processes of abrasion and reconstruction of strata, and the mountain mass has developed itself, until we have, as the final result, a series of approximately parallel ranges, showing in their structure the complicated nature of the processes by which they have been formed.

This method of growth by lateral aggregation is most admirably exemplified in the Cordilleras of North America. In this complex of chains, we have, first, the granitic and gneissic nucleus or basis, which is the floor on which all the stratified formations have been laid down, and from whose ruins the bulk of the materials have come for building up the ranges. This ancient nucleus is, in places, low down and concealed by heavy masses of stratified formations; in other regions raised into lofty crests, possibly the highest of the whole series. The stratified deposits, which have been formed from this nucleus, have been, from time to time, folded, upheaved, and invaded by eruptive rocks, whose distribution, however, has been very irregular.

On the western or oceanic side the disturbances have been most extensive. Here the upturnings and crushings of the strata have taken place on the grandest scale, and new axes of elevation have been formed at successive geological epochs, the close of the Jurassic and of the Miocene tertiary being two of the most important of these. On the eastern or Rocky Mountain side no great folding or metamorphism of the rocks occurred after the close of the Azoic period ; but a gradual elevation of the whole mass of strata took place, the larger portion of which was during the Tertiary epoch. By this uplift the unaltered cretaceous rocks were raised to an elevation in places greater than ten thousand feet above the present sea-level. This rise of the land continued until the most recent geological times, or almost down to the present day ; but how much of the difference in elevation between the land and sea is due to actual positive uplift, and how much to a sinking of the ocean, we have at present scarcely any means of judging. At all events there were, on this side of the Cordilleras, almost no local disturbances or foldings of the sedimentary rocks, which still lie upon each other in regular sequence, dipping at a low angle from the central crystalline masses everywhere, except just at the line of junction of the two formations, where, for a distance of a few thousand feet at right angles to their trend, the stratified formations, from Silurian to Carboniferous, are turned up on edge in the most wonderful manner, and sometimes completely overthrown, so as to dip towards the mountains, but not metamorphosed or rendered crystalline in structure. Neither were these disturbances attended, to any considerable extent, by outbursts of volcanic or eruptive material ; while on the western side of the continent these occurred on the grandest scale.

An examination of all that has been published with regard to the geology of the Andes indicates that when this mighty chain of mountains comes to be thoroughly studied, there will be many analogies discovered between them and the North American Cordilleras. Some such could already be indicated if space permitted ; but, as yet, no careful section has ever been made across the South American ranges by any trained stratigraphical geologist.

In the case of the Appalachians, we have to do with a chain of mountains which has no crystalline centre or axis, and which consists, at least through a great portion of its length, of a pile of detrital materials, distinctly stratified, all belonging to the Palæozoic epoch, scantily provided with fossils, but separable into a number of well-marked groups by the aid of lithological characters. These groups have much their greatest development towards the northeast and southwest, and they dip in general towards the west or northwest, so that in going in that direction we rise on to more recent strata. Proceeding westerly, moreover, we find the plications, which are well marked on the eastern edge of the chain, gradually disappearing; while at the same time the groups of strata are found to be made up of finer materials and to be gradually thinning out, thus indicating a greater distance from the source from which the detritus of which they are made up was derived. Hence we can hardly fail to draw the inference that this source was somewhere to the east of the range, and that the region from which the plicating force proceeded is also to be sought for on that side. If this be the case, then it seems probable that there must have been a high range of crystalline rocks on the eastern borders of the Appalachians, for there is no other conceivable source of supply which would satisfy the required conditions. The detritus of which the rocks of this range are made up came then from a higher region, which has since disappeared. It must have subsided, and this subsidence was, as we conceive, the cause of the plication of the beds which had been formed on its western slopes, these beds having been elevated and crumpled or flexed as the mass exterior to them was gradually sinking.

Subsidence, then, we regard as the chief cause of the plication of strata; but it is not the sinking of the stratified mass itself which is the principal effective agent in bringing about its folding. There can be no plication, to any appreciable extent, without an actual shortening of the plicated strata, and this can only come from a lateral thrust, such as would be exerted by a subsiding mass upon a region exterior to it. Hence if we find the newer strata on the flanks of an older central nucleus compressed together by folding, we are justified in presuming that it is the subsidence of the latter which has given rise to the plication of the adjacent lower region. If the subsiding higher area be of comparatively large dimensions, there will be a tendency to produce elevation to a certain extent on each side.

The folding of the newer strata along the base of the Alps, and of the Jura even, has repeatedly been explained by successive upheavals of the Alpine masses; indeed, these have been taken for granted by most geologists, without any attempt to investigate the manner in which the assumed cause could bring about any such result. It is certainly clear enough that elevation of the central mass would produce a lengthening rather than a shortening of the base on which rest the strata which are uplifted, and that this is something quite the opposite of what is required to cause plication. From a careful study of the stratigraphical geology of the Sub-Himalayan ranges, Mr. Medlicott, of the India Survey, was led to the conclusion that the peculiar position of the rocks of which those mountains are made up could only be accounted for on the theory of a subsidence of the central mass, and the same idea has been applied by him to explain the contortions of the tertiary beds on the flanks of the Alps, as well as to plications of stratified rocks in general.* And so far as this geologist has developed his ideas on this subject, they are identical with those formed by us in studying the mountain systems of North America.

We have thus endeavored to give an idea of the progress making by geologists in getting towards a solution of some of the principal problems of orography. The subject is a very comprehensive and difficult one, and it is far from easy to treat it in a popular manner. It is evident that much remains to be done in this line of research, and that it is desirable that chemists and physicists should lend a helping hand ; but the burden of the work must fall on the geologists, and one important step will have been made when it is clearly recognized that geological facts must be allowed to have more weight than chemical theories, and that a large experience in the field is a necessary prerequisite to valuable theorizing.

---

* See Quarterly Journal of the Geological Society of London, XXIV. 34.